DEFEATING DARWINISM
by Opening Minds

Phillip E. Johnson

InterVarsity Press
Downers Grove, Illinois

InterVarsity Press® is the book-publishing division of InterVarsity Christian Fellowship®, a student movement active on campus at hundreds of universities, colleges and schools of nursing in the United States of America, and a member movement of the International Fellowship of Evangelical Students. For information about local and regional activities, write Public Relations Dept., InterVarsity Christian Fellowship, 6400 Schroeder Rd., P.O. Box 7895, Madison, WI 53707-7895.

Scripture quotations, unless otherwise noted, are from the New Revised Standard Version of the Bible, copyright 1989 by the Division of Christian Education of the National Council of the Churches of Christ in the USA. Used by permission. All rights reserved.

Cover photograph: Louis Psihoyos

ISBN 0-8308-1360-8 (pbk.)
ISBN 0-8308-1362-4 (cl.)

Printed in the United States of America ⊖

Library of Congress Cataloging-in-Publication Data

Johnson, Phillip E., 1940-
   Defeating darwinism by opening minds/Phillip E. Johnson.
      p.   cm.
   Includes bibliographical references (p.   ).
   ISBN 0-8308-1360-8 (pbk; alk. paper)
   ISBN 0-8308-1362-4 (cloth; alk. paper)
   1. Evolution (Biology)   I. Title.
QH366.2.J655   1997
576.8'2—DC21   97-12916
                                                                                      CIP

21	20	19	18	17	16	15	14	13	12	11	10	9	8	7	6	5	4	3	2	1
14	13	12	11	10	09	08	07	06	05	04	03	02	01	00	99	98	97			

To Roberta and Howard,
who understood "the wedge"
because they love
the Truth

Introduction

This book grew out of two conversations. The first was in spring 1996, with friends from InterVarsity Press, my usual publisher. The Press was ready for me to do another book, but I wasn't sure I was ready. I had done a book *(Reason in the Balance)* a year earlier, I had just come back from a long lecture tour, I was immersed again in law-school work, and I had a lot of magazine writing to do. I wanted to take my time before beginning another project as demanding as a book.

As we talked, however, it became clear that there *was* one book I needed to write very soon. I had taken on the scientific evidence for Darwinian evolution in *Darwin on Trial* in 1991, and I had gone into the philosophical, moral and educational consequences of Darwinism in *Reason in the Balance* in 1995. Both books were successful and helped to open up a renewed public debate about whether Darwinism is really true. Both went into considerable detail about scientific and intellectual subjects, however, and a lot of readers who needed to know the basic message found them heavy going.

There was clearly a need for a short book aimed at a different audience, one not quite so familiar with university-level subjects. In particular, I wanted to write for late teens—high-school juniors and seniors and beginning college undergraduates, along with the parents and teachers of such young people.

These young people need to take advantage of the wonderful educational opportunities our society offers, but they also need to protect themselves against the indoctrination in naturalism that so often accompanies education. Textbooks and other educational materials today take evolutionary naturalism for granted, and thus assume the wrong answer to the most important question we face: Is there a God who created us and cares about what we do? Young people need to be prepared for the indoctrination, and for that they need to know some things that the public schools aren't allowed to teach them. That's the main job of this book, and everybody I've talked to seems to agree that it's a job that needs to be done.

That brings me to the second conversation, which occurred in the faculty club of my own university. I remarked to one of my senior Berkeley colleagues that the scientific community was baffled at its failure to convince the general public to believe in evolution. Despite massive educational efforts—including a pitch for evolution on every public television program that deals with nature—the state of public opinion hasn't changed much in the last thirty years. Polls show that under 10 percent of the American public believes in the official scientific orthodoxy, which is that humans (and other living things) were created by a materialistic evolutionary process in which God played no part. The remaining 90 percent is more or less evenly divided between biblical creationists and theistic evolutionists (who think evolution was God-guided). Why won't the people believe what the evolutionary scientists tell them science has discovered?

My colleague commented, "It's just that the people don't understand the theory."

"Oh no," I blurted out in answer. "The people understand the theory better than the scientists do."

My colleague looked at me as if he were trying to decide whether I was joking or insane, and we let the matter drop. As I thought over what I had said, however, I realized how true it was.

My experience speaking and debating on this topic at universities has taught me that scientists, and professors in general, are often confused about evolution. They may know a lot of details, but they don't understand the basics. The professors typically think that evolution from molecule to man is a single process that can be illustrated by dog breeding or finch-beak variations, that fossil evidence confirms the Darwinian process of step-by-step change, that monkeys can type *Hamlet* if they are aided by a mechanism akin to natural selection, and that science isn't saying anything about religion when it says that we were created by a purposeless material process.

All those beliefs are egregiously false, as I will explain in the chapters to come. Many ordinary people are also confused about these subjects, of course, but they do tend to grasp one big truth that the professional intellectuals usually seem incapable of seeing. The people suspect that what is being presented to them as "scientific fact" consists largely of an ideology that goes far beyond the scientific evidence. That is why they are so resistant to it. If high-schoolers need a good high-school education in how to think about evolution, professors and senior scientists seem to need it just as badly.

That's what this book aims to give—a good high-school education in how to think about evolution. It's for high-schoolers, college students, parents, teachers, youth workers, pastors and also scientists whose education didn't encourage them to take a skeptical look at the claims of Darwinian theory. There isn't much scientific detail in the book, or much advanced philosophy. I've covered the science and the philosophy in my earlier books, and refer readers to the relevant chapters as appropriate. I'll also refer in the research notes to some helpful teaching materials that are available on the World Wide Web. As additional materials become available, they'll be announced at the Access Research Network Web site (http://www.arn.org/arn). I'm sure it won't be long before we will have a first-rate Internet site available where teachers and parents can exchange insights about teaching techniques and materials.

As this book's title indicates, understanding evolution is mainly a matter of opening minds, of freeing people to think about it as they would other important subjects. All it really takes is precise definitions and good thinking habits. The skills you'll develop in learning to understand evolution will come in handy for a lot of other things too. Actually, you'll find out that they are the same skills that scientists like Carl Sagan have advocated all along. It's just that we are going to apply those skills to evolution, a subject that has for too long been protected from critical thinking by law and academic custom.

Emilio's Letter

Three Common Mistakes

A student from a European university posted an e-mail message on a public Internet forum in order to explain how creationists and evolutionists can make peace. I've changed the spelling and grammar a bit to conform to standard American usage, but in substance the message is just as this young man, whom I will call Emilio, wrote it:

> I've been a Christian and a creationist all my life, fiercely against evolutionism until I started my Biology course at the University and began learning about evolution. Guess what: I am still a creationist and now I am also an evolutionist! It has become clear to me that the first chapter of Genesis is an allegory (if not check how it states that there was morning and evening before the creation of the Sun and Moon—an impossibility), and once we accept this there is no reason why God could not have created all there is in as many million years as you wish.
>
> If God created time and space, then he is outside of it and therefore is not affected by it—time has no meaning to God! I

believe that God created the laws of physics, and therefore everything that results from such laws is God's creation. To say that the species *evolved* does not deny God's act of creation. Quite the opposite: evolution is the science that studies how God created the species.

Furthermore, evolutionism and creationism cannot be put in the same category, as one is of science, of the rational, and the other is of faith, of the supernatural.

I am a Christian, and it offends me to see that Christians are being viewed as lamebrains just because some well-intentioned but ignorant brothers of mine try to discuss such matters without scientific knowledge—please stop, you do more harm than good.

Emilio's statement brought a rueful smile to my face, because I have heard the same reasoning so many times from American Christians, and from agnostics who have figured out how to get Christians to adopt agnostic ways of thinking. Of course Emilio is doing his best to cope with a difficult situation, but he has committed the three basic mistakes that have done more than anything to disable Christians and other theists from talking sense about God to the world.

Mistake Number 1: It's Only About Length of Time

First, Emilio has trivialized the conflict between evolution and creation, portraying it as merely a dispute over whether the word *day* in the book of Genesis can be interpreted figuratively rather than literally. His logic is that if the "days" of Genesis are really a poetic way of describing long geological ages, then "evolution" is merely God's chosen method of creating, and one can without difficulty be both an evolutionist and a creationist.

Like many mistakes, this one contains grains of truth. St. Augustine wrote many centuries ago that God is outside of time and the Creator of time, and certainly God could use a gradual process of creation over millions of years if that is what he wanted to do. Emilio is right that we shouldn't refuse to consider genuine scientific possibilities just because we insist on reading Genesis more literally than perhaps its author intended.

Unfortunately, this much-too-easy solution to the problem rests on a misunderstanding of what contemporary scientists mean by that word *evolution*. If they meant only a gradual process of God-guided creation, then Emilio might be on the right track. A God-guided process is *not* what modern science educators mean by "evolution," however. They are absolutely insistent that evolution is an *unguided* and mindless process, and that our existence is therefore a fluke rather than a planned outcome.

For example, the 1995 official Position Statement of the American National Association of Biology Teachers (hereafter NABT) accurately states the general understanding of major science organizations and educators:

> The diversity of life on earth is the outcome of evolution: an unsupervised, impersonal, unpredictable and natural process of temporal descent with genetic modification that is affected by natural selection, chance, historical contingencies and changing environments.

Or, in the words of the famous evolutionist George Gaylord Simpson, "Man is the result of a purposeless and natural process that did not have him in mind."

I will explain in subsequent chapters why the biologists insist that evolution must be unsupervised and why God's purposes are not listed among the things that might have affected evolution. For now I will just say that this claim is not one they can afford to abandon, because their whole approach is founded on *naturalism*,[1] which is the doctrine that "nature is all there is."

If nature is all there is, then nature had to have the ability to do its own creating. Darwinian evolution is a theory about how nature might have done this, without assistance from a super-

---

[1] *Naturalism* and *materialism* mean essentially the same thing for present purposes, and so I use the terms interchangeably. Naturalism means that nature is all there is; materialism means that matter (i.e., the fundamental particles that make up both matter and energy) is all there is. Because evolutionary naturalists insist that nature is made up of

natural Creator. That is why "evolution" in the Darwinian sense is by definition mindless and godless. Pretending otherwise is an evasion of the conflict, not a resolution of it. Yet many Christian theologians and educators take this evasive approach because they are hoping to find an easy way to avoid coming to grips with a very difficult problem.

That's *mistake number one:* Emilio is kidding himself about what "evolution" means. It doesn't mean God-guided, gradual creation. It means unguided, purposeless change. The Darwinian theory doesn't just say that God created slowly. It says that naturalistic evolution is the creator, and God had nothing to do with it.

Mistake Number 2: God Made the Laws and Then Retired
What I have said so far is fairly well known because the scientific authorities have spelled it out again and again. I suspect Emilio understands this at some level, because he also proposes a different way of reconciling evolution and creation, and thus makes his second mistake. Even if evolution itself was mindless, God could be still be the author of the ultimate laws of nature by which evolution operates. Thus we have Emilio's second attempt: God made the laws of physics and chemistry, and evolution follows those laws. Therefore God is ultimately the Creator of everything, even if evolution was, as the Darwinists say, unsupervised and purposeless.

The notion that God is a remote First Cause who establishes the scientific laws and thereafter leaves nature to its own devices is called *deism.* Deism is different from *theism,* which implies a God who takes an active supervisory role in the world—like the God of the Bible. When Darwinists say that their theory does not

---

those particles, there is no difference between naturalism and materialism.

In other contexts, however, the terms may have different meanings. *Materialism* is sometimes used to mean greedy for material possessions, as in "he who dies with the most toys wins." *Naturalism* also has quite different meanings in other contexts, such as art and literary criticism. These other meanings are irrelevant for our purposes.

deny "the existence of God" and claim that they are saying nothing about "religion," they usually mean that they are willing to allow deism as a possibility for people who are unwilling to give up God altogether. Many evolutionary naturalists see no harm in making this concession, because a God who confines his activity to the ultimate beginning of time is unimportant to human lives.

The important question is not whether God "exists"; it is whether God cares about us, and whether we need to care about God's purposes. Deism answers no to these questions. For that reason even George Gaylord Simpson found deism to be perfectly consistent with his Darwinian doctrine that our true creator is a purposeless material system.

So that's *mistake number two*. Emilio is willing to exchange the Creator God of the Bible for the lifeless First Cause of deism. It's like trading real gold for counterfeit money.

Mistake Number 3: Giving Away the Realm of Reason
That deism isn't Christianity is pretty well known also, and again Emilio seems to have some awareness of this. So he makes a third attempt, and in the process rediscovers an error that many prominent theologians and philosophers have made before him. The Bible says that "in the beginning, God created," and "in the beginning was the Word." The kind of science Emilio is worried about says that "in the beginning were the particles" and "creation required no preexisting mind or purpose." One account starts with God, the other with matter. They seem absolutely opposed to each other. But could it be that the opposition is only apparent, the result of comparing unlike things as if they were similar?

Consider this example. Artist says: "The Grand Canyon is sublimely beautiful." Scientist says: "It's a big old hole in the ground." These statements may seem contradictory on a superficial reading, but of course they are both perfectly true. The Grand Canyon *is* a big old hole in the ground. It's also sublimely beautiful. The superficial contradiction disappears when we understand that an artistic statement about beauty deals with a different realm from a purely descriptive statement about the bare facts that the Grand Canyon is old and deep.

Maybe the same thing is true about apparent contradictions between religious and scientific statements. Maybe the statement "In the beginning God created" is like the artistic statement about the Grand Canyon. It's about how we perceive the meaning and beauty of the world. Maybe the scientific claims about mutation and natural selection are like the statement that the Grand Canyon is a big old hole in the ground. They are about bare physical facts, and hence they do not really contradict the higher-level statements about meaning, purpose, beauty and God.

In that case the scientific facts do not matter—whatever they are. Emilio can proclaim with relish a complete naturalism in science and insist that it makes no difference to faith. He might explain his position in words like these: "Yes, the diversity and complexity of life are the result of evolution. Yes, evolution is a blind, unsupervised and unintelligent process. Yes, we humans are the result of a purposeless and natural process that did not have us in mind. Isn't it wonderful that science [reason] has discovered all this *knowledge?* Of course none of this scientific knowledge contradicts my religious *belief* that God is our maker, because science is known to us by reason and religion is a matter of faith."

The "faith versus reason" (and belief versus knowledge) mistake is very seldom stated that clearly. Clear, simple statements tend to arouse our common sense, which tells us that Emilio is trying to ride two horses that are going in opposite directions. Rather, this third mistake thrives in the obscurity provided by big words and lengthy academic books. The third mistake is not as simple as the first two. It is a sophisticated mistake, and hence it has an irresistible attraction to intellectuals who are looking for a way to convince themselves that there is no need to deal with the conflict between theism and scientific naturalism. Fortunately, Emilio states the mistake artlessly, so we can more easily understand what is wrong with it.

Exactly what *is* wrong with the idea that religious statements belong to the realm of faith while scientific statements to the realm of reason? Let's consider how Emilio's professors and fellow students understand this distinction. They know that people believe in all kinds of things: astrology, spirit guides that communicate with dead people, space aliens that kidnap humans for bizarre scientific experiments, and the God of the Bible. From a scientific naturalist viewpoint, all such beliefs are about equally irrational. Any of them can be justified by an appeal to "faith." Against such an appeal, rational argument and evidence are helpless.

Here's a familiar example. Some small children believe that Santa Claus brings the presents they receive at Christmas. They believe it on authority ("Mom and Dad told me") and because it is a satisfying myth, one that makes them think they are loved by some great being beyond their family experience.

Then one day a child hears from a slightly older friend that Santa is not real. At first she regards this as an unholy lie, but the friend talks our little girl into performing a scientific investigation. She keeps herself awake after bedtime somehow and tiptoes to the top of the stairs to watch her parents putting the presents into the Christmas stockings.

What will she believe? If she says that reason tells her one thing and faith tells her another, we expect her to choose reason as she grows up. If she continues to believe in Santa as the years go by, she is refusing to come to terms with reality. We may treat her with kindness and even think she is cute, but we will not take her seriously. If she goes on believing in Santa as an adult, we will conclude that she is insane.

Is the God of the Bible, or the Jesus who rose from the dead, an adult version of Santa Claus? One can defend a myth of that kind by saying that it gives people comfort or allows them to think that death is not the end of everything. The myth may even be socially useful, provided it is kept within bounds. Even atheists sometimes worry about the moral consequences for society if religion disappears altogether. These are reasons for treating the story with a certain patronizing respect and for allowing believers to keep their illusions if they must.

Christian theists can buy the tolerance of most agnostics if they retreat to the "faith escape." The price is that they tacitly agree not to dispute the naturalistic doctrine that God, like Santa Claus and Zeus, is a product of human culture. In consequence, for intellectual purposes Christianity ranks among what some call the "higher superstitions," meaning the kinds of irrational belief that are relatively respectable in polite society. Christian beliefs are studied in religious studies programs alongside primitive myths as subjective belief systems that are not based on scientific knowledge.

So that's *mistake number three.* I said it was a sophisticated mistake, and one can argue that it is not so much a mistake as a rational defensive strategy born of desperation. In fact I wouldn't blame Emilio if he were to break into the conversation at this point and reply in these words: "What exactly is it that you want me to do? As a boy I was taught to believe in a literal Genesis story that contradicts modern science at just about every point. I can't defend that story now that I'm in the university. I'd be buried under a mountain of evidence coming from experts who know far more about every subject than I do. I wouldn't have confidence in my own arguments, and I'd be labeled a crank. My career in science would be over before I even got started. Unless I'm just going to give up and become an agnostic (like most of the other students), I don't see what else I can do but rely on these defensive measures you call 'mistakes.' "

Do You Have a Better Idea?
It's the right question to ask, and the answer is that I do. The rest of this book will be devoted to explaining that idea. Here for a start is what Emilio has to do if he is to master the challenge of evolution instead of having it master him.

First, Emilio has to stop seeing this issue as a conflict between the Bible and science, in which the supposed problem is to decide between two complete stories and to believe uncritically in one or the other. As we shall see, evolutionary naturalists rely on a cultural stereotype to shut off all criticism of their philosophy. The stereotype portrays all opponents as extreme Genesis literalists who reject the evidence of science for purely religious reasons. As long as the conflict is perceived this way, the grave scientific defects in the ruling theory, and the philosophical bias that sustains it, can be effectively concealed from view.

Second, Emilio needs to focus above all on what I call the "blind watchmaker thesis." This is the Darwinian claim that God was not necessary for biological creation, because the impersonal material forces of genetic mutation and natural selection can and did produce all the fantastic complexity of living organisms. The scientific evidence is strongly *against* the blind watchmaker thesis, and therefore strongly against the NABT's claim that a purposeless material process is our true creator.

Third, Emilio has to learn that "science" as defined in our culture has a philosophical bias that needs to be exposed. On the one hand, science is *empirical.* This means that scientists rely on experiments, observations and calculations to develop theories and test them. On the other hand, contemporary science is *naturalistic and materialistic* in philosophy. What this means is that materialist explanations for all phenomena are assumed to exist. And what that means is that the NABT's definition of evolution as an unsupervised process is simply true by definition—*regardless of the evidence!* It is a waste of time to argue about the evidence if one side has already won the argument by defining the terms.

I'll explain all these points in detail in subsequent chapters. First I should explain why it is so important not only for Christians, but for everyone who loves the truth and wants to see it prevail, to understand how to deal rationally with the tyranny of Darwinism. I'll use U.S. culture as a case in point; though the situation takes a different form in other societies, similar errors in thinking are widespread, as Emilio's letter makes clear.

The United States of America at the beginning of the twenty-first century is in a most peculiar situation. This country is the world leader both in science and in setting cultural trends, but it is unusual among industrialized nations in its refusal to abandon the concept that God is our Creator. Most Americans say they believe in God, and many of them really seem to mean it. Opinion polls indicate that no more than one out of ten Americans accepts "evolution" in the NABT sense—as a purposeless material system that accounts for all of life's history. Of the approximately 90 percent who attribute our existence to God, almost half are "creationists" in the sense that they reject an evolutionary explanation of human origins altogether. The other half are like Emilio; they mistakenly think they have resolved the problem by viewing evolution as a God-guided system of gradual creation.

Although most Americans believe in a Creator, the intellectual culture is totally dominated by naturalism—even in many colleges and seminaries that are formally religious. That God is an invention of human culture is taken for granted not only in the natural sciences but throughout the many departments of our universities, including "religious studies" departments. Many theologians and leaders of mainstream Christian denominations accept this situation or even endorse it, denouncing "creationists" for being ignorant troublemakers—as Emilio is learning to do.

In short, the intellectual elite in America believe that God is dead. In consequence they think that reason starts with the assumption that nature is all there is and that a mindless evolutionary process absolutely *must* be our true creator. The common people aren't so sure of that, and some of them are very sure that God is alive.

I agree with the common people. If we are right, the consequences are very, very important. The ruling naturalists know that too, although they may deny it. That is why they are so determined to define words like *evolution* and *science* in such a way that naturalism is true by definition.

I therefore put the following simple proposition on the table for discussion: *God is our true Creator.* I am not speaking of a God who is known only by faith and is invisible to reason, or who acted undetectably behind some naturalistic evolutionary process that was to all appearances mindless and purposeless. That kind of talk is about the human imagination, not the reality of God. I speak of a God who acted openly and who left his fingerprints all over the evidence. Does such a God really exist, or is he a fantasy like Santa Claus?

That is the subject of this book.

Inherit the Wind

The Play's the Thing

*A*fter almost every lecture I give, some person—usually a parent—asks me for advice about how to come across as a reasonable person when speaking up at a school-board meeting against the dogmatic teaching of Darwinian evolution. People who only want unbiased, honest science education that sticks to the evidence are bewildered by the reception they get when they try to make their case. Their specific points are brushed aside, and they are dismissed out of hand as religious fanatics. The newspapers report that "creationists" are once again trying to censor science education because it offends their religious beliefs. Why is it so hard for reasoned criticism of biased teaching to get a hearing?

The answer to that question begins with a Jerome Lawrence and Robert E. Lee play called *Inherit the Wind,* which was made into a movie in 1960 starring Spencer Tracy, Gene Kelly and Frederic March. You can rent the movie at any video store with a "classics" section, and I urge you to do so and watch it carefully after reading this chapter. The play is a fictionalized treatment of the "Scopes Trial" of 1925, the legendary courtroom confrontation in Tennessee over the teaching of evolution. *Inherit the Wind* is a masterpiece of propaganda, promoting a stereotype of the public debate about creation and evolution that gives all virtue and intelligence to the Darwinists. The play did not create the stereotype, but it presented it in the form of a powerful story that sticks in the minds of journalists, scientists and intellectuals generally.

If you speak out about the teaching of evolution at a public hearing, audience and reporters will be placing your words in the context of *Inherit the Wind*. Whether you know it or not, you are playing a role in a play. The question is, which role in the story will be yours?

The Story of the Play

A handsome young science teacher named Bert Cates, dedicated to his students and his teaching, is jailed for violating a state law against the teaching of evolution. Bert is in love with Rachel Brown, also a teacher and the daughter of the Reverend Jeremiah Brown, the most powerful of the local ministers. Reverend Brown is a vicious bigot with no redeeming qualities whatsoever, whose practice of Christian ministry seems to be limited to cursing people like Bert and threatening them with damnation. Rachel herself is a conformist; although she adores Bert, she continually urges him to stop making trouble for himself by speaking out against the community's religious prejudices.

The trial of Bert Cates becomes America's first media circus when Matthew Harrison Brady volunteers to be the prosecutor. Brady, a former presidential candidate, has become an antievolution crusader in his declining years. As the town of Hillsboro is preparing to give Brady a hero's welcome, journalist E. K. Hornbeck arrives from Baltimore. Hornbeck is a familiar movie character: the hard-boiled reporter who makes sarcastic comments about events on stage, a bit like the chorus in an ancient Greek drama. The townspeople provide him with many opportunities to exercise his wit, as they display their ignorance and vulgarity while mindlessly singing choruses of "Give Me That Old-Time Religion."

Brady eventually arrives, makes a phony-sounding speech, eats picnic food like the glutton he is, and generally shows himself to be a pompous old fool. He is also sneaky. After meeting the Reverend Brown and learning that Rachel Brown is friendly with Bert Cates, he induces the gullible Rachel to confide in him about ideas Bert has shared with her in confidence. Brady treacherously intends to use these against Bert in court, and even to call Rachel as a prosecution witness against her future husband.

The Brady welcoming banquet is interrupted by the news that the famous Henry Drummond is coming to be the defense lawyer. Drummond is another familiar movie character: the fearless advocate who fights for justice against seemingly hopeless odds. As the trial begins we see him trying to counter the religious prejudice of the community and the court. His every witticism strikes home, just as every feeble attempt by Brady to score a point backfires. If the deck in Hillsboro is stacked heavily against Drummond, the deck in Hollywood is stacked just as heavily in his favor. This black-and-white morality play could not be starker: all intelligence and goodness are on the side of Drummond and Cates, all folly and malice belong to Brady and Brown.

Although the defense is pure in mind and heart, it has an impossible legal position. Bert admits that he taught evolution, and that is what the law forbids. The prosecution proves its case by making some reluctant students testify that they were taught evolution. Brady unnecessarily supplements this evidence by forcing Rachel to testify to Bert's dangerous opinions, of which the most dangerous is this: "God created man in his own image, and man, being a gentleman, returned the compliment." Rachel's testimony has no legal significance, but its dramatic purpose is to underscore Bert's kindness and decency. He forgives Rachel and puts himself at risk by forbidding Drummond to upset her with cross-examination. Drummond has brought several scientific and theological experts to Hillsboro to testify that Darwinism is scientifically valid and no danger to a properly rational religion. The judge rules the expert testimony inadmissible, thus leaving the defense temporarily at a loss.

Drummond brilliantly saves the situation by calling his adversary Brady to the witness stand as an expert on the Bible. The judge correctly points out that this testimony is also irrelevant to the question whether Bert violated the law, but Brady is so conceited that he insists on taking the stand to show he can defeat the unbeliever. Drummond skillfully takes advantage of Brady's overconfidence. After some preliminary sparring about details like Jonah and the whale, Drummond stuns Brady by pointing out that the biblical patriarchs did their "begetting" by sexual intercourse. Apparently Brady had not previously thought of this embarrassing but undeniable fact, and he blurts out that the Bible calls sex "original sin." The dramatic point, of course, is that Bible believers are killjoys and prudes who want to abolish sex.

Eventually a rattled Brady concedes that since the first day of creation occurred before the sun existed, it might have been longer than twenty-four hours. Drummond seizes on this concession to demolish Brady's confidence, and gets Brady to talk such obvious nonsense that even his supporters laugh at him. The day ends in a spectacular moral victory for the defense.

None of this has anything to do with the legal issue, so the jury returns a guilty verdict the next day anyway. The town fathers are sufficiently embarrassed by the fiasco, however, that they pressure the judge to impose a nominal fine in the hope that this will end the publicity. Bert refuses to pay the fine and vows to go on speaking up for truth and freedom. Brady desperately attempts to retrieve the situation with another speech and is so upset by his own incoherent rant that he has a stroke and dies on the spot. Rachel tells Bert that she has decided to start thinking for herself, which in the context of the play seems to mean that she will accept Bert's way of thinking instead of her father's. (I can't help wondering whether her new independence of mind will have unexpected consequences, and whether Bert will ever have any second thoughts about having encouraged it.) The two lovers decide to leave town and get married. Love and reason thus overcome prejudice and bigotry.

As the play ends, Drummond is left alone on the stage court-room with his reflections. According to the stage directions, he picks up a copy of Darwin's *Origin of Species* and a copy of the Bible, "balancing them thoughtfully, as if his hands were scales. He half-smiles, half-shrugs." Then he jams the two books to-gether into his briefcase. The symbolism tells us that the Bible and Darwin can balance each other, if we allow Henry Drummond to do the balancing. It is roughly the line of reasoning that we saw Emilio accepting at the beginning of the previous chapter.

The Scopes Trial: What Really Happened

As the authors of *Inherit the Wind* admit in their preface, the play is not history. That is an understatement. The real Scopes trial was not a serious criminal prosecution but a symbolic confron-tation engineered to put the town of Dayton, Tennessee, on the map. The Tennessee legislature had funded a new science edu-cation program and, to reassure the public that science would not be used to discredit religion, had included as a symbolic measure a clause forbidding the teaching of evolution. The governor who signed the bill, realizing that any prosecution would be an embarrassment, predicted that the law would not be enforced. The American Civil Liberties Union wanted a test case, however, and advertised for a teacher willing to be a nominal defendant in a staged prosecution. Local boosters in Dayton took up the offer in the hope that the mock trial would be good for business. The volunteer defendant, John T. Scopes, was a physical education teacher who taught biology briefly as a substitute. He was never in danger of going to jail.

Local prosecutors fell in with the scheme and obligingly obtained an indictment against Scopes, respectfully declining the ACLU's offer to pay the costs of the prosecution. The trial got out of hand and became a media circus when William Jennings Bryan volunteered to speak for the prosecution and Clarence Darrow volunteered to be the defense lawyer. Darrow, fresh from a sensational murder trial in Chicago, was also nationally famous as an agnostic lecturer. Bryan, a three-time Democratic presidential candidate, was no reactionary but a progressive politician who had led political battles to protect working people and farmers from the excesses of big business. His reasons for opposing Darwinism appealed to many liberals and socialists in his day, as they still would. Bryan had seen Darwinism used in America to justify unrestrained capitalism, and in Germany to justify the brutal militarism that led to World War I.

Like Bryan, Clarence Darrow was a friend of labor unions and an opponent of unrestrained capitalism. He was also a materialist and a determinist who defended his clients by denying that they possessed free will. Darrow did not want to balance the Bible with evolutionary science; he wanted to get rid of religion and replace it with science and agnostic philosophy. On the other hand, Bryan truly was a scientific ignoramus, and the wily Darrow really did make a fool of him. If Darrow had wanted, he probably could just as easily have made the leading evolutionary scientists of the day look foolish. For example, some of these scientists confidently cited the fraudulent Piltdown Man and the tooth of "Nebraska Man" (which turned out to be from a kind of pig) as proof of human evolution. If Bryan was confused about the evidence for evolution, he had a lot of respectable company.

What the Play Means

I won't go any further into the discrepancies between the play and history, because the play has had so much impact that its story is more important than what really happened. The play is not primarily about a single event; it is about the modernist understanding of freedom.

Once upon a time, the story says, the world was ruled by cruel religious oppressors called Christians, similar to the wicked stepmother and stepsisters in "Cinderella," who tried to prevent people from thinking and from marrying their true love. Liberation from this oppression came via Darwin, who taught us that our real creator was a natural process that leaves human reason free to make up new rules whenever we want. Most modernist intellectuals interpret the story that way, and of course a liberated Cinderella is not likely to give the wicked stepmother another chance to enslave her. Whatever the stepmother says, Cinderella knows who she is and what she wants to do.

Read that way, *Inherit the Wind* is a bitter attack upon Christianity, or at least the conservative Christianity that considers the Bible to be in some sense a reliable historical record. The rationalists have all the good lines and all the virtues. Brady and Brown are a combination of folly, pride and malice, and their followers are so many mindless puppets. One would suppose from the play that Christianity has no program other than to teach hatred. At the surface level the play is a smear, although it smears an acceptable target and hence is considered suitable for use in public schools.

Just how ugly the smear is came home to me the first time I saw the movie, in a theater next to Harvard University (at a time when I would have called myself an agnostic). The demonstrative student audience freely jeered at the rubes of Hillsboro, whooped with delight at every wisecrack from Hornbeck or Drummond, and reveled in Brady's humiliation. It occurred to me that the Harvard students were reacting much like the worst of the Hillsboro citizens in the movie. They thought they were showing how smart they were by aping the prejudices of their teachers and by being cruel to the ghost of William Jennings Bryan—who was probably a much better man than any of them.

Maybe Hillsboro isn't just Dayton, Tennessee. Maybe sometimes it's Harvard, or Berkeley.

The Story Told Another Way

That memory has stayed with me, and shows that there may be more than one way to interpret the play. I've told the bare bones of the story literally; now let me retell the story at a different level, with just a tad of artistic license.

A brilliant young teacher develops a following because he has exciting ideas that open up a new way of life. His friends and students love him, but the ruling elders of his community hate the very thought of him. These elders are themselves cruel hypocrites who pile up burdens on the people and do not lift a finger to help them. The elders rule the people by fear and are themselves ruled by fear. They substitute dogmas and empty rituals for the true teaching they once knew, which commands truth and love as its first principles.

The elders want to destroy the teacher who threatens their control over the people, but his behavior and character are so exemplary that they can find no fault to justify condemning him. They plan to entrap him by convincing one of his closest friends to betray him. Eventually they are able to arrange a rigged legal proceeding and get a guilty verdict. Their victory is empty, however. The teacher wins even when he apparently loses, and he sums up his teaching in these words: "You shall know the truth, and the truth shall make you free."

Does that story sound familiar? Of course Bert Cates is not Jesus, although the play does portray him as virtually sinless. It would be more accurate to say that the authors aimed to give Cates and Drummond the virtues of Cinderella and Socrates. My point is that even this most seemingly antibiblical of dramas achieves its moral effect by borrowing elements from the gospel, which is the good news of how we can be delivered from the power of sin. Sin has power over us in many ways, and one of them is through the mind control practiced by fearful and hypocritical religious authorities. The independent mind that overthrows such oppressive power is good news for everyone but the oppressor.

Inherit the Wind is therefore probably truer than its authors knew. There is nothing wrong with its basic story of liberation. That story itself becomes a vehicle of oppression, however, when it invites the people with power to cast themselves as the liberators. It's like the dictators of the former Soviet Union calling themselves the champions of the poor working man. Whatever may have been the case a long time ago, by the time the movie was made Bert Cates and Henry Drummond were the ones with the power to shut other people up.

Owning the Microphone
My summary of *Inherit the Wind* left out two events that I now want to bring into the picture. When Henry Drummond is humiliating Matthew Harrison Brady on the witness stand, he accuses Brady of setting himself up as God by presuming to suppress freedom of thought in others. Drummond warns Brady that someday the power may be in other hands, saying, "Suppose Mr. Cates had enough influence and lung power to railroad through the State Legislature a law that only *Darwin* should be taught in the schools!"

That possibility may have seemed remote in Hillsboro, but of course it is exactly what happened later. The real story of the Scopes trial is that the stereotype it promoted helped the Darwinists capture the power of the law, and they have since used the law to prevent other people from thinking independently. By labeling any fundamental dissent from Darwinism as "religion," they are able to ban criticism of the official evolution story from public education far more effectively than the teaching of evolution was banned from Tennessee schools in the 1920s.

But how was this reversal accomplished in a voting democracy? Given that a majority of Americans still believe that God is our Creator, how have the Darwinists been able to obtain so much influence and lung power?

The play answers that question too. In the final scene of *Inherit the Wind,* when the jury returns to the courtroom to deliver its verdict, a character identified as "Radio Man" appears in the courtroom, carrying a large microphone. He explains to the judge that the microphone is connected by direct wire to Station WGN in Chicago. Radio Man proceeds to report directly to the public on the proceedings as they happen. Brady, famed for decades as an orator with a huge voice, attempts to speak into the microphone but can't master the technique. During Brady's final tirade the radio program director decides that his speech has become boring, and Radio Man breaks in to announce that the station will return to the Chicago studio for some music. The stage directions describe this as Brady's "final indignity," and it brings on his fatal stroke.

The microphone (that is, the news media) can nullify Brady's power by (in effect) outshouting him. But does this development imply liberation, or a new form of control that will be more oppressive than the old one? There is only one microphone in that courtroom, and whoever decides when to turn it on or off controls what the world will learn about the trial. That is why what happened in the real-life Scopes trial hardly matters; the writers and producers of *Inherit the Wind* owned the microphone, making their interpretation far more important than the reality. Bert Cates didn't have enough lung power to make law in Hillsboro, but his successors had enough microphone power to take over the law at the national level.

When the creation-evolution conflict is replayed in our own media-dominated times, the microphone-owners of the media get to decide who plays the heroes and who plays the villains. What this has meant for decades is that Darwinists—who are now the legal and political power holders—nonetheless appear before the microphone as Bert Cates or Henry Drummond. The defenders of creation are assigned the role of Brady or of the despicable Reverend Brown. No matter what happens in the real courtroom, or the real schoolroom, the microphones keep telling the same old story.

This has very practical consequences. I have found it practically impossible, for example, to get newspapers to acknowledge that there are scientific problems with Darwinism that are quite independent of what anybody thinks about the Bible. A reporter may seem to get the point during an interview, but after the story goes through the editors it almost always comes back with the same formula: creationists are trying to substitute Genesis for the science textbook. Scientific journals follow the same practice. That Matthew Harrison Brady may have valid scientific points to make just isn't in the script.

Danny Phillips

Occasionally a dissenter from Darwinism threatens to take over the role of Bert Cates. Here is one example: Danny Phillips was a fifteen-year-old high-school junior in the Denver area who thought for himself. His class was assigned to watch a *Nova* program, produced with government funds for National Public Television, which stated the usual evolutionary story as fact. Its story went something like this: "The first organized form of primitive life was a tiny protozoan. . . . From these one-celled organisms evolved all life on earth."

Science education today encourages students to memorize that sort of naturalistic doctrine and repeat it on a test as fact. Because Danny has a special interest in truth, however, and because his father is pastor of a church that has an interest in questioning evolutionary naturalism, Danny knew that this claim of molecule-to-man evolution goes far beyond the scientific evidence. So he wrote a lengthy paper criticizing the *Nova* program as propaganda. School administrators at first agreed that Danny had a point, and they tentatively decided to withdraw the *Nova* program from the curriculum. That set off a media firestorm.

Of course Danny was making a reasonable point. The doctrine

that some known process of evolution turned a protozoan into a human is a philosophical assumption, not something that can be confirmed by experiment or by historical studies of the fossil record. But the fact that administrators seriously considered any dissent from evolutionary naturalism infuriated the Darwinists, who flooded the city's newspapers with their letters. Some of the letters were so venomous that the editorial page editor of the *Denver Post* admitted that her liberal faith had been shaken. She wrote that "these defenders of intellectual freedom behaved, in fact, just like a bunch of conservative Christians. Their's was a different kind of fundamentalism, but no less dogmatic and no less intolerant."

In other words, at least one editor wasn't sure who was playing what role in the revival of *Inherit the Wind*. When Danny's story appeared on CBS television a little later, however, an experienced Darwinist debater named Eugenie Scott was careful to cast Danny as the opponent of learning. She argued, "If Danny Phillips doesn't want to learn evolution, . . . that's his own business. But his views should not prevail for eighty thousand students who need to learn evolution to be educated." When evolution is the subject, questioning whether the official story is true is enough to make you an enemy of education.

This manufactured image of a high-school sophomore censoring science education replaced the real Danny Phillips on national television, just as *Inherit the Wind* replaced the real Scopes trial. What Danny said when he got a chance to speak for himself was reported only in a local paper. He said, "Students' minds are to be kept open and not limited by a set of beliefs." That is exactly the right line to take, and Danny had for a moment a partial success in getting past the microphone owners. The CBS network and the Denver school board decided against Danny in the end—but then, the Hillsboro jury also decided against Bert Cates. All they inherited was the wind.

An Uphill Battle

In subsequent chapters I'll be explaining how some of us are working to make it possible for evolution to be treated like other issues, where criticism of the official story can get a fair hearing. It is an uphill battle, because Darwinists can use their control of the microphone to cast their opponents as religious dogmatists regardless of what the opponents are actually saying. If critics object to the teaching of philosophical doctrines as scientific facts, the microphones say that they are trying to prevent students from learning. If critics attempt to tell the other side of the story and bring out evidence that the textbooks ignore, they are accused of trying to insert religion into the science curriculum in violation of the Constitution. The rule of the microphone is "Heads we win, tails you lose."

It isn't easy to win a game played by those rules, but there is a way to do it. The first step is to learn how to detect baloney.

3

Tuning Up Your
Baloney Detector

*T*he late astronomer and science popularizer Carl Sagan worried that an epidemic of irrationality is loose in the world. Millions read astrology columns in the newspapers. People who ought to know better are taken in by faith-healing scams or believe that aliens in flying saucers kidnap people to perform scientific experiments on them. What is just as bad, wrote Sagan, is that most Americans continue to believe that they were created by God despite everything that he and other prominent scientists have done to persuade them that nature is all there is. What we need to protect ourselves from such false beliefs, Sagan writes in his book *The Demon-Haunted World,* is a well-equipped "baloney detector kit." A baloney detector is simply a good grasp of logical reasoning and investigative procedure.

Carl Sagan and I would agree about how to describe the principles of baloney detecting in general. We would disagree only about where the detectors are to be pointed, and especially about whether we should ever suspect the presence of baloney in claims made by the official scientific establishment. Sagan was that familiar figure in the modern scientific culture, the selective skeptic. His debunking skills were directed against the con artists and eccentrics who work on the margins of society, but he was an unquestioning true believer in the pronouncements of mainstream science about subjects like evolution.

Let me describe the varieties of baloney that every baloney detection kit should be equipped to recognize. They are basically the same ones Sagan listed, but I'll apply them to some examples of my own.

Selective Use of Evidence

There is a whole lot of evidence out there, and even a false theory is likely to be supported by some of it. That is the main reason it is so important to keep a debate open to dissenting points of view: one side shouldn't be allowed to just ignore evidence it finds inconvenient.

I see this point continually illustrated in debates over evolution. For example, textbooks and museum exhibits highlight fossils that can be interpreted as possible transitional forms between major groups—fossils that are actually quite few in number. They rarely inform the public about the far greater mass of contrary evidence, such as the absence of ancestors for the major animal groups that appear in the Cambrian explosion. I have written elsewhere about the "Hard Facts Wall" museum exhibit in San Francisco, which goes so far as to supply imaginary common ancestors for the animal groups, thus leading unwary visitors to think the ancestors have actually been found. Visitors to the museum at first take the exhibit at face value; after I explain it to them, they are astonished that a reputable museum would commit such a deception. But the museum curators are not consciously dishonest; they are true believers who are just trying too hard to help the public to get to the "right" answer. Without dissenters, such misrepresentations would go uncorrected.

So don't be impressed by claims that specific fossils, like the bird/reptile *Archaeopteryx* and the hominid Lucy, prove the theory of evolution. All such fossils are at most possible ancestors of living groups (like modern birds and humans), and a lot of interpretation is involved in classifying them. Insist on asking the right question: Does the fossil evidence, considered as a whole and without bias, tend to confirm the predictions of Darwinian theory?

Appeals to Authority

Nothing is true just because some big shot says it is true. Sagan tells us that "in science there are no authorities; at most, there are experts." Of course the experts sometimes get the idea that they *are* authorities and that what they say must be right just because they said it. The best check on this human tendency to be dogmatic is the test of experiment. A really good experimental test can call everybody's bluff.

I like to illustrate this point by telling a fictionalized version of the *Challenger* space shuttle disaster. Imagine that the launch of the shuttle has been delayed several times by bad weather or technical problems, and there is a lot of political pressure to go ahead and not to delay any further. Despite some misgivings because conditions are not ideal, top scientists and NASA administrators agree that the launch should proceed. A lowly student intern upsets this happy consensus by saying that the temperature is too cold and so the seals won't work and the rocket engine will explode. Nobody will listen to her because she has no status, although she has worked out the calculations carefully. If the top people go ahead with the launch and the engine explodes as the intern said it would, there's no doubt who was right and who was wrong. Reputations and status don't count for anything against the test of experiment.

Science would never go far wrong if direct and conclusive experimental tests were always possible. Unfortunately, sometimes only very limited tests can be made, and not all the tests will agree. In that case, scientific conclusions may be based on the opinion of the experts, who arrive at their judgment by a process of debate and negotiation. The result that comes out may depend more on who has the power than on who has the right answer. That's the difference between politics and science.

To illustrate, let's suppose that the *Challenger* launch was just a practice simulation and the rocket engines weren't actually ignited. Suppose Congress is fed up with delays and cost overruns, and top NASA executives are afraid their budget will be cut if they don't report a successful launch. That means a lot of people down the line will lose their jobs or research funds. So the top scientists and managers want very badly to report that the launch was a success. When the experts get together to write that report, who is going to pay attention to the student intern who did all the careful calculations? I'm not suggesting that the experts will lie. Rather, they will be under enormous pressure to reason their way somehow to the conclusion that she must have made a mistake somewhere. If the intern insists on pressing her point too far, she will endanger her own future in science. Nobody wants to hire a troublemaker.

Ad Hominem Arguments

A person with the wrong motives may have the right answer. Be careful about ad hominem arguments, which attack the person making the argument instead of the argument itself. (*Ad hominem* is Latin for "to the man.") Attacking somebody as a creationist, or an atheist, is often a way of distracting attention from valid arguments that person has to offer.

On the other hand, it is not necessarily irrelevant or unfair to point out that a person has a bias. Again, the problem is not so much that people might lie as that we all have a tendency to believe what we want to believe. If a man argues that secondhand cigarette smoke isn't hazardous to your health, nobody thinks it unfair to point out that he owns a cigarette company or that he has smoked heavily for years and doesn't want to think that he may have endangered the health of his family. His bias is relevant, but it doesn't necessarily mean he is wrong. That depends on the evidence.

In almost every disputed matter there is a problem of bias on both sides, and it's legitimate to bring this out. Bible believers may be reluctant to credit evidence that seems to contradict some passage in the Bible, and atheists may be reluctant to credit evidence that seems to suggest that natural selection can't do all Darwin claimed for it. Business owners don't like to believe facts that may hurt their business, and zealots for consumer protection may exaggerate the conclusions of a single study that confirms their worst suspicions about business. Scientists may be biased in favor of theories that make their work important and hence tend to increase their funding.

In this imperfect world an ad hominem argument sometimes performs the legitimate function of showing that a person has a bias and hence that his or her arguments should be examined carefully. The argument is misused if it does more than that, causing us to ignore worthwhile arguments because of what we think of the person making them. The point is to recognize and acknowledge bias, and then get beyond it to evaluate the evidence fairly.

Straw Man Argument

A "straw man" argument distorts somebody's position in order to make it easier to attack. Creationists are particularly vulnerable to this kind of attack. That is so in part because some creationists really have made crazy arguments and in part because of the *Inherit the Wind* stereotype. Many Darwinists want to pretend that the only people who doubt their theory are the most extreme religious fundamentalists. They know how to win a debate when the issue is framed as "science versus the Bible," and so they want to keep the debate framed that way.

Contrariwise, Darwinists are in trouble when they have to present positive evidence that natural selection can create new

kinds of plants and animals from simple beginnings. Hence they
are constantly trying to divert the discussion away from the
scientific issues so that they can debate the straw man position
that we should close our eyes to scientific evidence if it seems to
contradict Genesis. One prominent science writer wrote to me
for months, never engaging the scientific issues but constantly
pestering me with questions about my interpretation of Genesis
("Did Adam have a navel?"). Obviously he was hoping to find a
straw man to ridicule.

Begging the Question
An argument is said to "beg the question" if it assumes an answer
to the very point that is in dispute. Here's a simple example:
  Question: Why should I believe the Bible?
  Answer: Because the Bible says so.
  Arguments defending Darwinism often seem to beg the question because they assume the point at issue, which is whether the
scientific evidence really does support the theory. Here's a
typical example:
  Question: What evidence proves that life evolved from nonliving molecules?
  Answer: Don't reject a scientific theory just because you have
a religious prejudice.
  The answer assumes the point in dispute, which is whether the
evidence for the chemical evolution of life is so overwhelming
that only a prejudiced person would be skeptical of it. Question-
begging arguments typically assume that science or reason is on
the arguer's side; then the person tries to put *you* in the position
of arguing against science and reason. If you let a straw-man
maker define the terms of the argument that way, you've lost
before you make your first point. Insist on a level playing field.

Lack of Testability
Learn to distinguish between theories that put themselves at
risk—that is, invite testing by observation or experiment—and

theories that can't be shown to be either true or false. Everything scientists say isn't necessarily scientific, and some theories that come from eminent scientists may be as speculative as a theologian's musings about what heaven is like. Sagan gave the example of the "many worlds" hypothesis in quantum physics, which suggests that there may be many other universes other than the one we inhabit, so that every possible physical event has actually occurred somewhere. This could be true, but as there is no way to check out the existence of the other universes, it remains mere speculation. An even better example is Sagan's own statement at the beginning of his famous *Cosmos* television series: "The Cosmos is all there is, or ever was, or ever will be." What experiments can we perform to test that statement?

Either creation or evolution can be stated in both safe and risky forms. If I say I believe in creation on faith, no matter what the evidence is, then we can't test my belief by scientific observations or experiments. But if I say the evidence indicates that living organisms are necessarily the products of intelligent design and that life never could have emerged by purely natural means from a prebiotic soup of chemicals, my statement invites scientific testing. Theories of chemical and biological evolution aim to contradict my hypothesis of intelligent design, by showing that purposeless natural processes can do the creating by evolution. The question is whether they have been successful in doing this—that is, whether the theories have passed the experimental test or failed it.

Darwin's theory of evolution was originally stated in risky form. It predicted, for example, that fossil hunters would eventually find a great many transitional intermediates between the major groups (they didn't) and that animal breeders would succeed in creating distinct species (they didn't). Today the theory is usually stated in risk-free form. Naturalistic evolution is identified with science itself, and any alternative is automatically disqualified as "religion." This makes it impossible to hold a scientific debate over whether the theory is true (it's virtually true by definition), which explains why Darwinists tend to think that anyone who wants such a debate to occur must have a "hidden agenda." In other words, critics couldn't seriously be questioning whether the theory is true, so they must have some dishonest purpose in raising the question.

Vague Terms and Shifting Definitions

Make sure people don't mislead you by using vague terms that can suddenly take on a new meaning. In the creation-evolution debate, the key terms that are subject to manipulation are *science* and *evolution*. Everybody is in favor of science, and everybody also believes in evolution—when that term is defined broadly enough! But *science* has more than one definition, and so does *evolution*. Watch out for "bait and switch" tactics, by which you are led to agree with a harmless definition and then the term is used in a very different sense.

Here's an example of how you can be deceived: "You believe in dog breeding, don't you? Well, did you know that dog breeding is an example of evolution? Now that you know that, and have seen all those breeds of dogs for yourself, you realize that you actually *do* believe in evolution, don't you? Good. That's enough for today. Later on we'll tell you more about what evolution means." (It's going to mean that all living things are the accidental products of a purposeless universe.)

This is not a "straw man" example, by the way. Selective breeding of animals is a process guided by intelligence, and it produces only variations within the species; yet Darwinists from Charles Darwin himself to the more recent Richard Dawkins and Francis Crick have cited it as a powerful example of "evolution."

If somebody asks, "Do you believe in evolution?" the right reply is not "Yes" or "No." It is: "Precisely what do you mean by *evolution?*" My experience has been that the first definition I get will be so broad as to be indisputable—like "There has been change in the course of life's history." Later on a much more precise and controversial definition—like the one by the National Association of Biology Teachers I quoted in chapter one—will be substituted without notice.

That one word *evolution* can mean something so tiny it hardly matters, or so big it explains the whole history of the universe. Keep your baloney detector trained on that word. If it moves, zap it!

Original Sin

Finally, *watch out for the universal human tendency to believe what we want to believe.* I call this the "original sin" in science because it is the one big temptation that all the specific rules of baloney detecting are designed to protect us from.

Even top scientists have to guard against the temptation to believe what they want to believe. For one thing, their funding may depend on an experiment coming out "right," and so they may be tempted to accept too readily a preliminary test that gives the result they want. That is why scientists place so much importance on repeatable experiments, meaning experiments that give the same result when they are performed by other scientists who don't necessarily have the same reason to want a particular result. That is also why there is a connection between good science and democratic political values like freedom of thought and freedom of speech. Unpopular dissenters often insist on pointing out the facts that powerful people might prefer to ignore.

Trustworthy Experts

There's plenty more to be said about baloney detecting, but if you understand these basic points you are well on your way to becoming a good critical thinker. Rather than consider more refinements, we need now to consider a fundamental problem with the whole project of critical thinking. We can't possibly think out everything for ourselves all the time. Much of the time we have no alternative but to trust the experts. But how do we know whether we *can* trust them? The experts know more than we do, but they may also have an interest in persuading us to believe something that is in their own interests rather than our interests. They may give us what is popularly known as a "snow job."

Trustworthy experts are ones who understand their responsibility to give us their expertise without claiming to know more than they really do. Really trustworthy experts don't try to evade our baloney detectors, and even warn us to watch out for their own expert bias.

The best description I know of the qualities that make an expert trustworthy comes from the late great physicist Richard Feynman, one of the unquestioned heroes of modern science. If a teenager with a passion for science wanted to take one twentieth-century scientist as a model, he or she couldn't do much better than to pick Feynman. In his 1974 commencement speech at the California Institute of Technology. Feynman told the graduating students to cultivate

> a kind of scientific integrity, a principle of scientific thought that corresponds to a kind of utter honesty—a kind of leaning over backwards. For example, if you're doing an experiment, you should report everything that you think might make it invalid—not only what you think is right about it: other causes that could possibly explain your results; and things you thought of that you've eliminated by some other experiment, and how they worked—to make sure the other fellow can tell they have been eliminated. . . . In summary, the idea is to try to give all the information to help others to judge the value of your contribution; not just the information that leads to judgment in one particular direction or another.
>
> The first principle is that you must not fool yourself—and you are the easiest person to fool. So you have to be very careful about that. After you've not fooled yourself, it's easy not to fool other scientists. You just have to be honest in a conventional way after that.
>
> I would like to add something that's not essential to the science, but something I kind of believe, which is that you should not fool the laymen when you're talking as a scientist. . . . I'm

talking about a specific, extra type of integrity that is [more than] not lying, but bending over backwards to show how you're maybe wrong, that you ought to have when acting as a scientist. And this is our responsibility as scientists, certainly to other scientists, and I think to laymen.

I would like to think that when the graduating students of Caltech heard those inspiring words, they all stood up and shouted "Amen!" Maybe the students really did react that way, but (alas!) scientists who are not as scrupulous as Richard Feynman often employ very different principles when they deal with the public. They are afraid we will come to the wrong answers if we do our own thinking, and so they try to bluff and intimidate us.

Sagan's Bluff

Let's take Richard Feynman as our prime example of a truly scientific thinker and ask ourselves what he would say about the following statement by Carl Sagan. The quoted statement comes from Sagan's final book, *The Demon-Haunted World,* the same book where he urged us not to be impressed by invocations of authority and to insist on asking whether claims put forward in the name of science are really testable:

> I meet many people who are offended by evolution, who passionately prefer to be the personal handicraft of God than to arise by blind physical and chemical forces over aeons from slime. They also tend to be less than assiduous in exposing themselves to the evidence. Evidence has little to do with it. What they wish to be true, they believe is true. Only nine percent of Americans accept the central finding of modern biology that human beings (and all the other species) have slowly evolved by natural processes from a succession of more ancient beings with no divine intervention needed along the way.

Sagan here turns his baloney detector around. It's no longer a light to protect us from a snow job. It's a club to browbeat us into believing, against our better judgment, that humans arose by blind physical and chemical forces over eons from slime. (This central finding comes, mind you, from a scientific establishment that also insists that it isn't saying anything about God.) The statement has the form of critical thinking—it speaks of people who ignore evidence and believe what they want to believe—but there is no real attempt to reason. Is it really likely that 91 percent of the public disagrees with Sagan for no reason at all?

Let's consider two possibilities. One is that 91 percent of the public consists of ignorant people who ignore the evidence and just believe what they want to believe. On that assumption, democracy is a farce. We are like children who think we can set fires and not be burned. In that case we ought to be ruled by a scientific elite, who will protect us from the consequences of our folly. The other possibility is that the evolutionary naturalists are the ones who believe what they want to believe, and they are likewise the ones who are less than assiduous in exposing themselves to contrary evidence. Maybe Carl Sagan ignored Richard Feynman's warning: "The first principle is that you must not fool yourself—and you are the easiest person to fool."

Education or Indoctrination?

In a dictatorship, the dictator tells the people what they are supposed to believe. In a democracy, we try to educate citizens so they can reason for themselves. That doesn't mean we treat all answers as equally correct. The claim that 2 plus 2 equals 5 is not a dissenting opinion; it's a mistake! But we don't have to force people to believe the truths of arithmetic. If they are properly educated, they will accept them by reason. Democracy rests on the faith that ordinary people can be trusted with the powers of government if education teaches them to think rationally. This implies a democratic concept of education.

When good teachers are teaching more advanced problems in mathematics, or in other subjects, they love a student who will argue that the textbook answer isn't correct. The reason isn't so much that the textbook answer might be wrong, although that always is a possibility. The real reason is that people learn the truth best if they fully understand the objections to the truth. If I believe in evolution (or anything else) only because "Teacher says so," you could say I don't really believe in evolution. What I believe in is obedience to authority, and in letting "Teacher" do my thinking for me. A democratic education aims to produce citizens who can think for themselves. Carl Sagan would have agreed emphatically, and he would have said that unquestioning acceptance of the dictates of authority is the opposite of the kind of skeptical thinking science education ought to try to foster— except, of course, when it comes to evolutionary naturalism.

Given that only a small minority of Americans believe the central finding of biology—"that human beings (and all the other species) have slowly evolved by natural processes from a succession of more ancient beings with no divine intervention needed along the way"—how should our educational system deal with this important instance of disagreement between the experts and the people?

One way would be to treat the doubts of the people with respect, to bring them out in the open and to deal with them rationally. The opposite way is to tell the people that all doubts about naturalistic evolution are inherently absurd, that they should believe in the orthodox theory because the experts agree that it is correct, and that their silly misgivings will be allowed no hearing in public education.

American educators have chosen the second path, the path of Sagan's Bluff. I'll illustrate that with two examples that occurred in 1996.

The Lakewood Case
A high-school senior in Lakewood, a suburb of Cleveland, Ohio, wrote an editorial in the school paper in appreciation of physics teacher Mark Wisniewski. Wisniewski, a creationist, used a classroom exercise in which students were asked to think about how their own worldviews influence their interpretations of the de-

bate between creation science and the more orthodox scientific views of cosmology and biological evolution. The student later observed that Wisniewski "never stood on a soapbox and never made us feel like we were in Bible study. . . . The philosophical element is what made it special. [Wisniewski] wanted us to make up our own minds rather than spoonfeed us like other educators."

According to a commendably fair-minded article in the anti-creationist magazine *Skeptical Inquirer,* Wisniewski himself explained, "I tried to find something in the science arena [that would raise the worldview issues,] and the creation-evolution debate fits like a glove." Asked whether any other issue might illustrate his point as well without bringing religious debates into the classroom, Wisniewski argued that it is important that the dispute goes to core beliefs or the example wouldn't really hit home. He said his goal was to teach students how to interpret data on their own and not just "memorize and regurgitate the favorite interpretation of the teacher." He graded on how students supported their ideas and not on the ultimate answers they gave.

Unwittingly, the student got her favorite teacher in a peck of trouble by publicizing his teaching objectives and methods. No students in the class (or their parents) complained, but calls from out of town flooded the district's offices. Lawyers from the American Civil Liberties Union threatened the district with expensive litigation, and the district's own counsel advised administrators that they had better issue a directive forbidding teachers to raise the religious issues. Facing a lawsuit and a public controversy that would distract it from everything else, the district capitulated and ordered the teacher to stop.

The Response to Danny Phillips

At the end of the previous chapter I told the story of Danny Phillips, the Denver high-school student who startled his teachers by challenging teaching materials that present evolution as a fact. What he was challenging, of course, was the broad theory of evolution as defined by the National Association of Biology Teachers: "an unsupervised, impersonal, unpredictable, and natural process" that accounts for the entire history of life. Danny challenged evolutionary naturalism on two grounds: it is effectively a religious dogma, and it isn't supported by the weight of the scientific evidence. The school's administrators, impressed by Danny's arguments, initially ordered the offending film replaced by other teaching materials.

Danny's case ended like so many others; he lost because the power was on the other side. Self-styled "civil liberties lawyers" threatened to bring an expensive lawsuit, and the school board capitulated to them. Before that happened, however, Danny's challenge to evolutionary orthodoxy got a lot of newspaper and television coverage. Some of it was favorable, probably reflecting the natural sympathy many reporters feel for the student rebel who challenges the educational orthodoxy.

The uproar so upset science educators that they brought out a really big gun to squelch the high-school student. Bruce Alberts, president of the National Academy of Sciences, personally responded to Danny in an editorial published in the *Denver Post*. The NAS is the most prestigious organization of scientists in the United States, and so its president is effectively the official voice of the scientific establishment. Danny should have felt very honored to be engaged by so powerful an adversary.

Unfortunately, Alberts replied with the stock arguments that evolutionary naturalists use to silence discussion on this topic. He identified dissent from evolutionary naturalism with "religion," and hence with untestable speculation that science must disregard. As a clincher, he recommended that "those interested in understanding how science works may wish to read a recent book, *The Beak of the Finch,* by Jonathan Weiner, which describes new studies on the Galápagos Islands that confirm and elaborate on Darwin's original work. Evolution happens all around us."

Alberts was referring to studies which show that the average size of finch beaks on a particular island varies from year to year in response to environmental changes. (I discuss the Weiner book in chapter four of *Reason in the Balance*.) Anyone who has even the slightest acquaintance with the evolution-creation controversy would know that such minor variation is readily accepted by even the strictest biblical creationists. The evolution-creation controversy is not about minor variations but about how things like birds come into existence in the first place.

One of the truly bizarre things about our current cultural situation is that the leading figures of the scientific establishment seem genuinely amazed that the citizens do not accept finch-beak variation as proof of the claim that humans, like all animals and plants, are accidental products of a purposeless universe in which only material processes have operated from the beginning.

It's an absurd situation, isn't it? Educators aren't allowed to address the issues about which their students, and the general public, are most concerned. When teachers challenge students to think about how their worldviews affect their understanding of the creation-evolution controversy, so-called civil liberties lawyers censor the teaching by threatening to bring a lawsuit that the school district can't afford to defend. The president of the National Academy of Sciences writes an essay so simplistic that it insults the intelligence of a well-informed high-school student. He urges a bright high-school student not to think for himself but to trust the findings of a research community that thinks it can settle the question of our origins by defining finch-beak variation as "evolution."

How did the scientists get themselves into such a mess? It has to do with the way Darwinists think, and how they define *science*.

A Real Education in Evolution

*A*popular teacher encourages young people to raise the big issues and think for themselves, and gets in trouble for it. A bright young student takes a stand for freedom of thought, and runs smack into a wall of official dogma. The authorities use the law to intimidate dissenters and try to discourage citizens from thinking for themselves about the evidence for evolution.

Where have we seen all this before? It's a replay of *Inherit the Wind*, of course, with the characters trading roles. The possibility that Henry Drummond raised in the play has come true. The Darwinists did get a law saying that only Darwinism may be taught it the schools, but they got it from the Supreme Court, not the legislature.

In a 1987 decision, the Supreme Court held unconstitutional a Louisiana state law that attempted to require balanced treatment for creation and evolution in the public school classroom. A state may not require, said the majority opinion, that the "religious viewpoint that a supernatural being created human-kind" be given fair treatment as an alternative to evolution in science classes. In context that meant that the opposite opin-ion—that humankind was created by a purposeless natural proc-ess that cares nothing about us—would be taught as unchallengeable fact.

Justice Antonin Scalia argued in dissent that the people of a state, "including those who are Christian fundamentalists, are quite entitled, as a secular matter, to have whatever scientific evidence there may be against evolution presented in their schools, just as Mr. Scopes [Bert Cates in the movie] was entitled to present whatever scientific evidence there was for it." The majority emphatically disagreed. Only Darwin may be taught in the schools.

The predictable result of this one-sided educational and legal regime is that evolution has become the focus of a culture war instead of a subject that can be discussed constructively in educational institutions or in the political realm of negotiation and compromise. The science educators teach the students that they were created by evolution and that evolution is a purposeless and unsupervised natural process. Of course those statements go far beyond the scientific evidence and state a religious posi-tion, but educators also insist with a straight face that they are not saying anything about religion or God. If they *were* addressing the subject of religion, they would have to allow the other side to be argued. Therefore they must not be addressing it.

When students ask intelligent questions like "Is this stuff really true?" teachers are encouraged or required not to take the questions seriously. Instead they put the students off with pub-lic-relations jargon about how the scientific enterprise is reliable and self-correcting. In California, for example, state curriculum guidelines advise teachers not to go into the merits of objections to evolution in class (where other students might be influenced) but to tell objecting students to take such questions up with their parents or a minister. When a teacher does try to take the objections seriously, the result is likely to be a lawsuit from the American Civil Liberties Union or People for the American Way, plus bad publicity in the press. School administrators understandably capitulate and tell teachers and students to stop making trouble. In short, Bert Cates and Henry Drummond have far surpassed their predecessors in using the tools of power to keep dissent from getting out of hand.

The situation is obviously unfair to the dissenters, but never mind that for now. I'm more concerned to point out to the scientific community how bad it is for science and for education.

Here is what I want to say to the scientists and educators: History has taught us that an established religion tends to fall into bad habits, and the same thing may be true when a scientific establishment starts to act like a governmental body with an official ideology to uphold. The price of having that kind of position is that you are tempted to protect your power and wealth by defending things you shouldn't be defending, with methods (like doubletalk and intimidating threats of legal action) that you shouldn't be using. These become bad habits, and they eventually lead you into massive hypocrisy and self-deception. When you preach baloney detecting as the essential tool of science but make students turn their baloney detectors off when they get to the really important questions of origins, you convict yourselves every day of hypocrisy. You also lose the ability to think critically about your own beliefs, and eventually you set yourself up for the kind of embarrassment that destroyed Matthew Harrison Brady.

There is only one cure. No matter how badly you want to bury the tough questions, you have to acknowledge that those questions really are too tough to be settled with misleading slogans like "Evolution is a fact" and "Science and religion are separate realms." You have to admit that people have reasons for objecting to the materialist philosophy you are presenting in the name of science. If you are going to be educators instead of dogmatists, you are going to have to start dealing honestly with those objections.

You need to turn your baloney detectors on yourselves. It hurts a lot at first, but eventually you will learn to enjoy it. Trust me—I've tried it!

Critical Thinking in Evolutionary Biology

Can we begin to treat evolution as a subject for education rather than for a culture war? Of course we can! If I were designing a curriculum for high-school or college students in evolution, I would build it around the same principles of baloney detecting we considered in the preceding chapter. Here are some of the things I would want students to learn.

1. *Learn to distinguish between what scientists assume and what they investigate.* Contemporary scientists don't investigate what the Supreme Court called "the religious viewpoint that a supernatural being created humankind [or anything else]." They disregard that possibility because they consider things supernatural to be outside of science. In other words, scientists start by assuming that naturalism is true, and they try to give purely natural explanations for everything, including our existence.

Because of that assumption, scientists do not really consider whether evolution (as distinguished from creation) is *true* or whether evolution might be guided by God. They assume that evolution is the only possibility and that it is unguided, because in their minds both special creation and guided evolution fall in the territory of religion, not science. They also assume that natural selection has great creative power, not because that power can be demonstrated but because there is no better naturalistic alternative.

Students should regard the neo-Darwinian theory of evolution, then, as merely the best naturalistic explanation of our existence that science can provide. Whether it is true is another question, and we cannot go into that question unless we are allowed to consider the possibility that a Creator exists.

Understanding the crucial role of philosophy in Darwinism is the key to understanding why the theory is so controversial, and why scientists want so badly to dodge the hard questions. Biologists have authority over questions of biology, but they have no authority to impose a philosophy on society. Once the public understands what they are doing, the biologists will lose their power to exclude dissent. That is why it is so important for them to insist that "evolution is a *fact*." Change that to "evolution is a philosophy," and the game is over.

Did creation require a Creator? You can assume a negative answer to that question on philosophical grounds, or you can treat it as a question of fact open to scientific investigation, but you can't legitimately do both. I would teach students to be distrustful of textbook authors, or other authorities, who try to have it both ways.

2. *Learn to use terms precisely and consistently. Evolution* is a term of many meanings, and the meanings have a way of changing without notice. Dog breeding and finch-beak variations are frequently cited as typical examples of evolution. So is the fact that all the differing races of humans descend from a single parent, or even that Americans today are larger on average than they were a century ago (due to better nutrition). If relatively minor variations like that were all evolution were about, there would be no controversy, and even the strictest biblical fundamentalists would be evolutionists.

Of course evolution is about a lot more than in-species variation. The important issue is whether the dog breeding and finch-beak examples fairly illustrate the process that created animals in the first place. Using the single term *evolution* to cover both the controversial and the uncontroversial aspects of evolution is a recipe for misunderstanding.

At a minimum students must learn to distinguish between microevolution (cyclical variation within the type, as in the finch-beak example) and macroevolution (the vaguely described process that supposedly creates innovations such as new complex organs or new body parts). Don't be impressed by claims that in a few borderline cases microevolution may have produced, or almost produced, new "species." The definition of "species" is flexible and sometimes means no more than "isolated breeding group." By such a definition a fruit fly that breeds in August rather than June may be considered a new species, although it remains a fruit fly. The question is how we get insects and other basic groups in the first place. Darwinists typically (but not always) claim that macroevolution is just microevolution continued over a very long time. The claim is very controversial, and students should learn why.

3. *Keep your eye on the mechanism of evolution; it's the all-important thing.* Some Darwinists distinguish between what they call the "fact of evolution" and "Darwin's particular mechanism." The "fact" usually just means that organisms have certain similarities, like the DNA genetic code, and are grouped in patterns (mammals, fish, insects and so on). This pattern of nature is uncontroversial. What's controversial is the cause of the pattern, and particularly whether that cause involves a Creator or only a purposeless material mechanism.

The problem with separating the fact from the mechanism is that a so-called fact of evolution doesn't have much scientific content without a testable mechanism for changing one kind of creature into something entirely different, and especially for building the extremely complex organs that all living things possess. Darwin knew this: it's the first major point he makes in *On the Origin of Species.* The pattern of organisms would provide "unsatisfactory" evidence for evolution, he argued, "until it could be shown how the innumerable species inhabiting this world have been modified, so as to acquire that perfection of structure which most justly excites our admiration."

Darwin's mechanism was natural selection. Today, despite many efforts to find an alternative, there still isn't really a competitor to the two-part Darwinian mechanism of random variation (mutation) and natural selection. Darwinists argue with each other about the relative importance of chance and selection, but some combination of these two elements is just about the only game in town.

Remember that the mechanism has to be able to design and build very complex structures like wings and eyes and brains. Remember also that it has to have done this reliably again and again. Despite offhand references in the literature to possible alternatives, Darwinian natural selection remains the only serious candidate for a mechanism that might be able to do the job.

That, by the way, explains why many Darwinists are reluctant to make a clear distinction between microevolution and macroevolution. They have evidence for a mechanism for minor variations, as illustrated by the finch-beak example, but have no distinct mechanism for the really creative kind of evolution, the kind that builds new body plans and new complex organs. Either macroevolution is just microevolution continued over a longer time, or it's a mysterious process with no known mechanism. A process like that isn't all that different from a miraculous or God-guided process, and it certainly wouldn't support those expansive philosophical statements about evolution being purposeless and undirected.

In my experience, the distinction between the fact of evolution and the neo-Darwinian theory always turns out to be just a debating gimmick to hide the problem with the mechanism from scrutiny. Once the "fact" is established, it turns out to include the necessary mechanism, which is mutation and selection.

Don't let anybody tell you that the mechanism is a mere detail; it's what the controversy is mainly about. When critics subject the mechanism to detailed criticism, Darwinists very quickly run out of evidence. That's when they want to substitute a vague "fact," which will later be inflated to include the whole theory. It's another example of bait and switch.

4. *Learn the difference between testing a theory against the evidence and using selected bits of evidence to support the theory.* I've long been fascinated by the conflicting messages Darwinists provide concerning the fossil evidence. On the one hand, they proudly point

to a small number of fossil finds that supposedly confirm the theory. These include the venerable bird/reptile *Archaeopteryx,* the "whale with feet" called *Ambulocetus,* the therapsids that supposedly link reptiles to mammals, and especially the hominids or ape-men, like the famous Lucy. These examples, all from vertebrate animals, are pressed very insistently on me in debates as proof of the "fact" of evolution and even of the Darwinian mechanism.

I am not as impressed by such examples as Darwinists think I should be, because I know that the fossil record overall is extremely disappointing to Darwinian expectations. One prime example is the "Cambrian explosion," where the basic animal groups all appear suddenly and without evidence of evolutionary ancestors. What is even more interesting is that the evidence for Darwinian macroevolutionary transformations is most conspicuously absent just where the fossil evidence is most plentiful—among marine invertebrates. (These animals are plentiful as fossils because they are so frequently covered in sediment upon death, whereas land animals are exposed to scavengers and to the elements.) If the theory were true, and if the correct explanation for the difficulty in finding ancestors were the incompleteness of the fossil record, then the evidence for macroevolutionary transitions would be most plentiful where the record is most complete.

Here is how Niles Eldredge, one of the world's leading experts on invertebrate fossils, describes the actual situation:

No wonder paleontologists shied away from evolution for so long. It never seems to happen. Assiduous collecting up cliff faces yields zigzags, minor oscillations, and the very occasional slight accumulation of change—over millions of years, at a rate too slow to account for all the prodigious change that has occurred in evolutionary history. When we do see the introduction of evolutionary novelty, it usually shows up with a bang, and often with no firm evidence that the fossils did not evolve elsewhere! Evolution cannot forever be going on somewhere else. Yet that's how the fossil record has struck many a forlorn paleontologist looking to learn something about evolution.

Eldredge also explains the pressures that could easily lead a forlorn paleontologist to construe a doubtful fossil as an ancestor or evolutionary transitional. Science takes for granted that the ancestors existed, and the transitions occurred, so scientists ought to be finding positive evidence if they expect to have successful careers. According to Eldredge, "the pressure for results, positive results, is enormous." This pressure is particularly great in the area of human evolution, where success in establishing a fossil as a human ancestor can turn an obscure paleontologist into a celebrity. Human evolution is also an area where the evidence is most subject to subjective interpretation, because ape and human bones are relatively similar. If you find an ape or human bone that's a bit unusual, can you construe it as a piece of a prehuman ancestor? If you can, and if the other experts will support you, your future may be a glorious one.[1]

In light of these pressures and temptations, how confident should we be that fossils of "human ancestors" are really what they purport to be? Could the wish be father to the thought, as it so often is?

To forestall outraged protests, I should emphasize that there is nothing cynical about asking these questions, nor do they imply that anybody is committing a deliberate fraud. Remember the wise

---

[1] The ever-changing story of human evolution took a strange new turn late in 1996, when geochronologists announced a study from Java indicating that three human species (*Homo erectus,* Neanderthals and modern humans) apparently coexisted on the earth as recently as thirty thousand years ago. The *New York Times* (December 13, 1996) front-page story reported, "Until a couple of decades ago, scientists conceived of the human lineage as a neat progression of one species to the next and generally thought it impossible that two species could have overlapped in place or time." It also observed, "It is not known how much contact the three species had, or if they could interbreed." If they could interbreed, then it would be more accurate to say that they were all a single species, *Homo sapiens.* Such huge areas of uncertainty support my view that general conclusions about evolution should not be drawn from the human fossil record, where the evidence is scanty and the temptation to subjectivity in interpretation is particularly great. Today's "fact" is likely to be tomorrow's discarded theory.

words of Richard Feynman: "The first principle is that you must not fool yourself—and you are the easiest person to fool." Think how easy it would be for ambitious fossil hunters to fool themselves, when the reward for doing so may be a cover story in *National Geographic* and a lifetime of research funding. Think how much pressure the other physical anthropologists are under to develop standards that will allow *some* fossils to be authenticated as human ancestors. A fossil field without fossils is a candidate for extinction.

Keeping all that in mind, why do you think such a high proportion of the fossils used to prove "evolution" come from this one specialty? Why do you think Niles Eldredge, a specialist in marine invertebrates, uses hominid examples rather than the vast record of fossil invertebrates to argue the case for evolution? If anybody tries to tell you that questions like these are improper (as they probably will), your baloney detector should blow a fuse. A scientist who objects to scientific testing is like a banker who doesn't want the books to be audited by independent accountants. View such people with suspicion.

5. *Learn the difference between intelligent and unintelligent causes.* This is a distinction that many otherwise capable scientists do not understand, because their materialist philosophy teaches them to disregard it. I'll illustrate the point with a couple of examples.

Tim Berra is a professor of zoology at Ohio State University. He wrote a book that was published by the Stanford University Press with the title *Evolution and the Myth of Creationism: A Basic Guide to the Facts in the Evolution Debate.* Berra's book has much the same purpose as this book. It aims to explain, for nonscientists, how good thinkers should view the conflict between evolution and creation. Here is Berra's explanation of "evolution," which comes illustrated with photographs of automobiles in the middle of the book:

> Everything evolves, in the sense of "descent with modification," whether it be government policy, religion, sports cars, or organisms. The revolutionary fiberglass Corvette evolved from more

mundane automotive ancestors in 1953. Other high points in the
Corvette's evolutionary refinement included the 1962 model, in
which the original 102-inch was shortened to 98 inches and the
new closed-coupe Stingray model was introduced; the 1968
model, the forerunner of today's Corvette morphology, which
emerged with removable roof panels; and the 1978 silver anni-
versary model, with fastback styling. Today's version continues
the stepwise refinements that have been accumulating since
1953. The point is that the Corvette evolved through a selection
process acting on variations that resulted in a series of transitional
forms and an endpoint rather distinct from the starting point. A
similar process shapes the evolution of organisms."

Of course, every one of those Corvettes was designed by engi-
neers. The Corvette sequence—like the sequence of Beetho-
ven's symphonies or the opinions of the United States Supreme
Court—does not illustrate naturalistic evolution at all. It illustrates how intelligent designers will typically achieve their purposes by adding variations to a basic design plan. Above all, such
sequences have no tendency whatever to support the claim that
there is no need for a Creator, since blind natural forces can do
the creating. On the contrary, they show that what biologists
present as proof of "evolution" or "common ancestry" is just as
likely to be evidence of common design.

I described the credentials of Professor Berra and named the
publisher so nobody could accuse me of attacking a "straw man."
A distinguished university press would not publish such a book
without obtaining professional reviews certifying that its scientific explanations were reliable. Evidently the reviewers saw nothing wrong with equating automotive engineering and biological
evolution. I am not surprised, because evolutionary biologists
typically do not understand that sequences resulting from variations on common design principles (as in the Corvette series)
point to the existence of common design, not its absence. I have
encountered this mistake so often in public debates that I have
given it a nickname: "Berra's Blunder."

A somewhat more sophisticated version of Berra's Blunder is

to confuse artificial (that is, intelligent) selection with natural selection. Francis Crick, who is a celebrated molecular biologist and a fervent scientific materialist, argued the case for Darwinism in these words:

> If you doubt the power of natural selection I urge you, to save your soul, to read Richard Dawkins' book *[The Blind Watchmaker]*. I think you will find it a revelation. Dawkins gives a nice argument to show how far the process of evolution can go in the time available to it. He points out that man, by selection, has produced an enormous variety of types of dog, such as Pekinese, bulldogs, and so on, in the space of only a few thousand years. Here "man" is the important factor in the environment, and it is his peculiar tastes that have produced (by selective breeding, not by "design") the freaks of nature we see preserved all around us as domestic dogs. Yet the time required to do this, on an evolutionary scale of hundreds of millions of years, is extraordinarily short. So we should not be surprised at the ever greater variety of creatures that natural selection has produced on this much larger time scale.

Was Crick aware that domestic animal breeding requires a preexisting, purposeful intelligence? He seems to have sensed it an one level, and then wished the ugly fact away by a verbal antithesis ("selective breeding, not . . . 'design' ").

Once again we see the truth of Feynman's warning: *the easiest person to fool is yourself.* Only a powerful unconscious need to overlook the truth could have allowed Crick to conceal from himself that animal breeders are intelligent agents, not blind natural forces. Breeders use expert skills to select just the variants they want, and they carefully protect their overspecialized breeds from the natural selection that would otherwise prevent such "freaks" from surviving to reproduce their own kind. Selective breeding is not the same thing as natural selection, or even analogous to it. It is intelligent design.

Critical Thinking Is Good for Religion Too
Every scientific materialist who reads this will understandably

want to ask: "Are you willing to apply baloney detecting to religion, as well as science?" The answer is (emphatically) *yes!* I can't think of a better way to introduce students to Christianity than to invite them to read the Gospels with care and to ask all the tough questions. I'm also not particularly worried about how they answer those questions the first time through. Dealing with the tough questions is a lifelong business, and the most important educational point is not to try to spoonfeed students with oversimplified answers that won't stand the tests of time and experience. Here are two examples of the kinds of issues I'd like young people to begin to think about.

6. *The problem of suffering.* One of the seeming advantages of Darwinism is that it makes it unnecessary to ask why God permits the innocent to suffer and (sometimes) the wicked to prosper. In a materialistic universe, moral arbitrariness is only to be expected. As Richard Dawkins puts it, "Nature is not interested one way or the other in suffering, unless it affects the survival of DNA." Some religious people actually like Darwinism because they think it gets God off the hook. If (for some reason) the divine plan involved creating by means of scientific laws, then God couldn't intervene to prevent suffering without spoiling his own grand scheme. I don't find that convincing, but it's clear that some Darwinists believe in their theory less because of the scientific evidence than because they have theological or philosophical objections to supernatural creation.

Of all the errors of scientific materialism, the silliest is that resolution of the National Academy of Sciences that religion and science are separate realms that should never be considered in the same context. On the contrary, evolutionary scientists are obsessed with the "God question," and the problem of suffering is one important aspect of that question.

I would tell students that none of the usual answers to the problem of suffering is entirely satisfactory. I'd want my students to have some familiarity with the classic treatments of the problem, especially the book of Job and the Grand Inquisitor section

of Feodor Dostoyevsky's *The Brothers Karamazov,* as well as a good
Christian apologetic like C. S. Lewis's *The Problem of Pain.* I'd want
them to read the Psalms and the Gospels with the problem fully
in mind, and think about whether and how the suffering and
resurrection of Jesus help with it. I'd want them to understand
that some of the appeal of Darwinism stems from classic philosophical objections to the doctrine that the world is governed by
a Creator who loves us and cares about what we do. Above all,
I'd want them to face the fact that if science has its unsolved
problems, so does religion. We all see through a glass darkly—
but what glass should we try to see through?

7. *The problem of faith.* One of the illusions of scientific materialism is its insistence that materialists don't have faith commitments. Faith is not something some people have and others
don't. Faith also isn't something opposed to reason. Faith is
something that everybody needs to get started in any direction,
and to keep going in the face of discouragement. Reason builds
on a foundation of faith.

For example, scientific materialists have faith that they will
eventually find a materialistic theory to explain the origin of life,
even though the experimental evidence may be pretty discour-
aging for now. Because they have faith in their theory, Darwinists
believe that common ancestors for the animal phyla once lived
on the earth, even though those ancestors can't be found. Niles
Eldredge calls himself a "knee-jerk neo-Darwinist" in spite of the
invertebrate fossil record—because he is convinced, on philosophical grounds, that the theory must be true. That's every bit
as much of a faith commitment as the belief of a young-earth
creationist that all radiometric dating must be wrong because it
contradicts the literal words of Genesis—and because it is a lot
easier to deal with the problem of suffering if pain and death
first entered the world after human beings had sinned.

Given that every position has its difficulties, where should we
put our faith? To use the words that Jesus taught us, what is the
foundation of solid rock, and what is the foundation of sand?

The Christian says that the rock is God, and we should trust in the goodness of God all the more when the presence of evil and suffering inclines us to doubt. The materialist says that the rock is matter, and that we should never move from an unshakable faith in science and materialism even when we begin to be discouraged by the difficulties of explaining all the things that do exist without allowing a role to a Creator.

Beginning a New Century—and a New Millennium

Whatever their faith commitments, good thinkers ought to be dissatisfied about the way things stand at the present time. The evidence that can survive baloney detecting isn't likely to satisfy either materialists or creationists. It seems for now as if new forms appeared mysteriously and by no known mechanism at various widely separated times in the earth's history. Maybe we'll be stuck with a mystery like that indefinitely, but I think it more likely that the twenty-first century will see a scientific revolution that will completely change our understanding of the history of life.

If I'm right about that, the chance to participate in discovering that new understanding should be a thrilling prospect for young people looking forward to a career in science. What makes science sound boring is the impression the books give that the important things have already been discovered and all there is left to do is fill in the details. Showing young people that there is a lot we don't know—and that we may even be dead wrong about some of the things we think we *do* know—is the way to fire their imaginations.

I don't know what new theories the future may bring, but I think I know where the revolution will start. It will start with the realization that life is not the product of mindless natural forces. Life was designed.

5

Intelligent Design

---

*G*eorge C. Williams is not as well known to the public as Richard Dawkins or Stephen Jay Gould, but he is one of the world's most respected and influential evolutionary biologists. He is best known for pioneering the "gene selection" version of Darwinism which was popularized by Dawkins in *The Selfish Gene*. Very briefly, gene selectionism says that natural selection selects genes, not whole organisms. However fit a plant or animal may be, in the end it dies and returns to the dust of the earth. What remains are its genes, encoded in DNA, because the genes were passed on to the next generations in the process of reproduction.

This means that genes (rather than bodies or minds) are the central actors in the evolutionary drama. The story of life then goes something like this.

The Story of Life, Starring Gene
In the beginning there was a naked gene that somehow evolved

from a chemical soup. This gene had two important properties: it could reproduce by copying itself, and it could engage in some sort of chemical activity analogous to eating. Mistakes were made in the copying process, and so the descendants of the first gene had varying capabilities. Those that were better at "eating," and especially reproducing, left more offspring than their less proficient sisters and cousins, and so the Darwinian process of natural selection could begin.

Eventually some genes learned to make bodies in a process we now call "embryonic development." (As applied to humans, that's the development of the baby in the mother's womb.) Those genes that could make bodies had a competitive advantage in the struggle for survival, all the more so as their body-building capabilities improved. A slogan humorously captures the basic idea: "A chicken is just an egg's way of making another egg." Gene selectionists talk and write as if genes can think and plan strategies for survival. They do not mean this literally, of course, but as a metaphor for a process that is directed only by the blind force of natural selection.

In brief, the gene selection theory posits that particular types of genes improve their own chances for survival by making, or improving, organisms that are themselves good at surviving and reproducing. Natural selection thus ensures that the world will be dominated by those types of genes that happen to be good at making plants and animals that are good at passing their own genes on to descendants.

Gene selectionism is an example of what philosophers call *reductionism.* Reductionists claim that everything, including our minds, can be "reduced" to its material base. For example, Dawkins has written that the discovery of the structure of DNA and its genetic code "has dealt the final, killing blow to the belief that living material is deeply distinct from nonliving material." Life is matter, and only matter. Dawkins does not flinch from applying this philosophy to human beings: "We are survival machines—robot vehicles blindly programmed to preserve the selfish molecules [of DNA] known as genes." The only purpose of life is DNA survival: a person is nothing more than DNA's way of making more DNA like itself. That's materialist reductionism as articulated by Richard Dawkins, today's most influential evolutionary biologist.

Information and the Word

Although George C. Williams did more than anyone to develop the gene selection theory, he seems to be having second thoughts about the underlying reductionism. In a 1994 book (supplemented here by an interview published in 1995) he endorsed the very different idea that life contains a distinct nonmaterial component called information. Because this subject is so important and controversial, I had better quote his exact words:

> Evolutionary biologists have failed to realize that they work with two more or less incommensurable domains: that of information and that of matter. . . . These two domains can never be brought together in any kind of the sense usually implied by the term "reductionism." . . . The gene is a package of information, not an object. The pattern of base pairs in a DNA molecule specifies the gene. But the DNA molecule is the medium, it's not the message. Maintaining this distinction between the medium and the message is absolutely indispensable to clarity of thought about evolution.
>
> Just the fact that fifteen years ago I started using a computer may have had something to do with my ideas here. The constant process of transferring information from one physical medium to another and then being able to recover the same information in the original medium brings home the separability of information and matter. In biology, when you're talking about things like genes and genotypes and gene pools, you're talking about information, not physical objective reality.

Williams uses the novel *Don Quixote* as his example of the matter-information duality. A computer operating system like Windows 95 would provide a similar example. A book or a computer program contains complex information recorded in matter, whether the matter be ink and paper or a silicon disk. The information can be switched from one medium to another, or even stored in the human brain. The content of the book or the computer program is not specified by the physical or chemical laws governing the medium. If it were, all books would be alike—or perhaps would differ according to the qualities of the ink and paper used to write them. In fact the content of the message is independent of the physical makeup of the medium. *Don Quixote* loses nothing of its meaning or literary quality if it is printed on the cheapest paper, and a trashy romance novel does not improve in quality if it is printed on expensive silk. The medium and the message are two entirely different kinds of things. As Williams explains,

> You can speak of galaxies and particles of dust in the same terms, because they both have mass and charge and length and width. You can't do that with information and matter. Information doesn't have mass or charge or length in millimeters. Likewise, matter doesn't have bytes. . . . This dearth of shared descriptors makes matter and information two separate domains of existence, which have to be discussed separately, in their own terms.

That way of describing reality brings to mind the biblical description of how the world began. The Gospel of John begins with the memorable statement that "in the beginning was the Word." That is exactly how we would describe the creation of a literary work, or a computer program, or a building. In the beginning was the concept and the working out of that concept in the mind of the author or designer. Thereafter the concept was recorded, or realized, in matter. Matter is important, but secondary. The Word (information) is not reducible to matter, and even precedes matter. If only matter existed in the beginning, then the first verse of the Gospel of John—and the worldview of the Bible—is false. In the beginning were the particles, and everything else came only from them. A reductionist understanding of the universe leaves no room for God, much less for the Word of God.

If everything came from matter, and if the information in living organisms is located in genes made of DNA, then it seems logical to suppose that DNA and life are virtually the same thing. Williams singled out Dawkins for criticism on this point, as one who "defines a replicator in a way that makes it a physical entity duplicating itself in a reproductive process," adding that Dawkins "was misled by the fact that genes are always identified with DNA." If Dawkins has been misled, however, it is not for some trivial reason. It is because highly complex information that is independent of matter implies an intelligent source that produced the information, and the main point of Darwinism for Dawkins is to eliminate that possibility from consideration.

To see why this is so, consider the crucial role of an author in producing the information in a book. Williams himself uses this analogy, and anybody who uses a word processor can see at once what he means.

A Book Isn't Just Ink and Paper

This book you are reading, like any other, contains information written on paper with ink. The information did not always take that physical form, however. Originally I wrote it on a word processor, and it existed only as an electronic file on a computer disk. I sent some completed chapters by e-mail to friends and colleagues for criticism. The information in each chapter was exactly the same whether it was recorded on paper or on a computer disk on in some fragmented and disembodied form as it moved over the links of the Internet.

Information is also stored by some poorly understood means in our brains. If all the copies of Shakespeare's plays were destroyed, nothing would be permanently lost. Actors who had learned the roles could easily re-create the texts from memory.

Such examples tell us that information is an entirely different kind of stuff from the physical medium in which it may temporarily be recorded. It would be absurd to try to explain the literary quality or meaning of a book as an emergent property of the physical qualities of its ink and paper. The message comes from an author; ink and paper are merely the media. Similarly, the information written in DNA is not the product of DNA. Where did the information come from? Who or what is the author?

Physical laws cannot be the answer to that question. These laws do produce some fairly complex structures, such as snowflakes and crystals. In such cases the laws produce the same structure over and over again, with chance variations. Repetitive order has a very low information content. The same laws that form the crystals prevent any more complex ordering from emerging, because they ensure that the same pattern will always be repeated according to the formula.

Similarly, we might create a sort of book by programming a computer with a formula, like this one: "Keep repeating the word *stuff* until the printer runs out of paper." A book written that way would be very boring to read, and it would never get more interesting even if we kept on printing forever. The only variety would come from an occasional typographical error; otherwise it would be just more "stuff" all the way to eternity.

If physical laws cannot provide the information in a book, could random chance do the job? I won't bore you with the math, but just about everybody (including Richard Dawkins) agrees that it is essentially impossible to produce a coherent book of average length by randomly combining letters, spaces and punctuation marks. Even a single sentence—like "In the beginning was the Word"—is extremely unlikely to come from pouring out a random mix of letters and spaces. As I said, that is undisputed. Some do say, however, that chance can do the job if it is combined with some principle of selection.

Berra's Blunder Again
Many people with underpowered baloney detectors have been

misled on this critical point by a common Darwinian application
of Berra's Blunder. It actually is possible to produce a written
text by supplying random letters—*if* some selector (like a computer program) preserves every letter that happens to end up in
the right place. Thus we can get "In the beginning was the Word"
if the program supplies random letters until an *I* happens to
appear in the first space, or a *d* in the final space, and so on.
Whenever a letter appears in the correct slot, the program
preserves it there, like the uncovered letters on the TV program
Wheel of Fortune. Very soon the spaces will all be filled with the
correct letters and we will have the whole sentence.

The whole thing seems absurdly easy—so easy that you ought
to smell a rat. With a fast enough computer generating thou-
sands of random letters a second, we can reproduce the whole
Bible in a matter of hours, plus the Chicago telephone directory
as a bonus. All we have to do is write the Bible and whatever else
we want into the computer's memory first, and have the computer preserve the desired letters in the right places until all the
spaces are filled. Richard Dawkins actually uses examples like
this to illustrate the creative power of natural selection, and his
readers apparently don't see that it's just a trick. If a computer
selection program can duplicate a library that easily, can't natural selection make an organism?[1]

You probably have spotted the trick already, but I'll explain it
just to make sure. Computer selection, like automobile design,
illustrates intelligent planning (authorship), not chance or survival of the fittest. It is just as if an author were writing the target

---

[1] A book review by the editor of a magazine called *Skeptic* provides a typical example of
Berra's Blunder. Dismissing the possibility of intelligent design in biology, the editor
comments: "Genetic mutations are chancy, but natural selection and the evolution of
complexity are not. Natural selection preserves the gains and eradicates the mistakes.
A monkey randomly typing will never produce *Hamlet;* but a monkey that learns, or a
computer system that holds all correctly sequenced letters and disregards the rest (a la
natural selection) will peck out 'TOBEORNOTTOBE' in a matter of minutes. Does this
happen at the cellular level? It does."
I am amused by self-styled "skeptics," who invariably seem able to believe the wildest
nonsense if it supports Darwinism.

phrase, except that the author has to wait a bit for the right letters to appear in the right spaces. The first letters to appear are meaningless, and the computer knows which ones to select only because it has the target text in its memory.

Natural selection, on the other hand, is supposed to be mindless and hence incapable of pursuing a distant goal. If natural selection could preserve a presently meaningless mutation because it might become useful later on when other new mutations occur, this would imply that evolution is a purposeful process, supervised by a preexisting mind. As we have seen, supervised evolution is a gradualist version of creationism. As materialists use the term, it is not evolution at all.

Let's Review What We Know

So far we have the following basic points.

First, life consists not just of matter (chemicals) but of matter and information.

Second, information is not reducible to matter, but is a different kind of "stuff" altogether. A theory of life thus has to explain not just the origin of the matter but also the independent origin of the information.

Third, complex, specified information of the kind found in a book or a biological cell cannot be produced either by chance or at the direction of physical and chemical laws. Attempts to prove that it can typically employ variations on Berra's Blunder.

With those general principles in mind, now let's go to the biology. Are organisms designed, or are they the products of unintelligent natural causes?

Opening the Black Boxes of Biology

To answer that question, we have to look beneath the surface of life to the biochemistry underneath. The biologists of the nineteenth and early twentieth centuries who established Darwinism and materialism as scientific orthodoxy knew little of biochemistry, and imagined the cell to be something rather simple that could just ooze itself up out of some chemical broth.

The term "black box" grew out of the efforts of scientists to expose medical hoaxes. A quack doctor might offer to cure whatever ails you by hooking you up to a mysterious black machine with all sorts of dials and switches on the cover, but nothing inside. More generally, any machine that does wonderful things by a mechanism nobody knows is called a black box. The computers on which we write are black boxes to most authors, because we have only the vaguest idea how they work. Without a knowledge of molecular biology, bodily functions like vision and blood clotting are black boxes. We know they work wonders, but we don't know *how* they work. Without that detailed knowledge, a biologist's notion of how (say) vision might evolve is as valueless as my speculations about how to build a computer.

In his book *Darwin's Black Box* molecular biologist Michael Behe explains that scientists have begun to open the black boxes of biology, and they have revealed a fantastically complex world of interacting proteins and enzymes underneath. Here, just to give a sample, is Behe's description of part of the molecular mechanism for vision:

> When light first strikes the retina a photon interacts with a molecule called 11-cis retinal, which rearranges within picoseconds to *trans*-retinal. (A picosecond is about the time it takes light to travel the breadth of a single human hair.) The change in the shape of the retinal molecule forces a change in the shape of the protein, rhodopsin, to which the retinal is tightly bound. The protein's metamorphosis alters its behavior. Now called metarhodopsin II, the protein sticks to another protein, called transducin. Before bumping into metarhodopsin II, transducin had tightly bound a small molecule called GDP. But when transducin interacts with metarhodopsin II, the GDP falls off, and a molecule called GTP binds to transducin. (GTP is closely related to, but critically different from, GDP.)

There is a lot more like that—but don't worry, I'm not going to inflict it on you. If you are interested in molecular biology, go read Behe's book. If not, all you need to understand is that molecular mechanisms are *irreducibly complex*. What this means is simply that they are made up of many parts that interact in complex ways, and all the parts need to work together. Any single part has no useful function unless all the other parts are also present. There is therefore no pathway of functional intermediate stages by which a Darwinian process could build such a system step by tiny step.

Molecular mechanisms, Behe says, are as obviously designed as a spaceship or a computer. You can't explain the origin of any biological capability (like vision) unless you can explain the origin of the molecular mechanisms that make it work. Evolutionary biologists have been able to pretend to know how complex biological systems originated only because they treated them as black boxes. Now that biochemists have opened the black boxes and seen what is inside, they know the Darwinian theory is just a story, not a scientific explanation.

The Attempt to Climb Mount Improbable

Behe published his book in 1996, the same year in which Richard Dawkins published *Climbing Mount Improbable*. The mountain of Dawkins's title is biological complexity, because Dawkins cheerfully acknowledges that plants and animals really are extremely complicated and that the analogy we have made to books and computers is valid. In his vivid words,

Physics books may be complicated, but . . . the objects and phenomena that a physics book describes are simpler than a single cell in the body of its author. And the author consists of trillions of those cells, many of them different from each other, organized with intricate architecture and precision-engineering into a working machine capable of writing a book. . . . Each nucleus . . . contains a digitally coded database larger, in information content, than all 30 volumes of the *Encyclopedia Britannica* put together. And this figure is for *each* cell, not all the cells of the body put together.

Dawkins rules out the possibility that such a database could be created all at once. To do that would require a prodigious mind and hence would amount to supernatural creation. Just as a mountain climber has to go up a mountain step by step, biological evolution has to go through a series of intermediate, functional steps in order to create each biological system—including the mind and body of the author of that physics book. Each step represents a random mutation, usually defined as a copying error in the reproduction of DNA. This means that the steps must be very small indeed, because mutations that are large enough to have visible effects on the organism are nearly always harmful or even fatal.

We may not see the intermediate steps today, but every Darwinist must believe that they once existed as actual living organisms. If such a thing as truly *irreducible* complexity actually exists, Dawkins concedes, then the functional intermediate steps could not have existed and Darwinism is not true.

Who is right, Dawkins or Behe? And is the argument about science or about philosophy? Behe says (and I agree) that the dispute is mainly philosophical. Science, he writes, is published in professional, peer-reviewed scientific journals. Behe's search of the professional journals reveals the absence of any serious efforts to lay out plausible, testable scenarios for the step-by-step evolution of molecular mechanisms. Such half-hearted attempts as exist are full of what scientists call "hand-waving." New molecular steps mysteriously "stand forth," or "emerge," or just "appear"—without any realistic mechanism. Molecular biologists don't even attempt to fill in the Darwinian theory with specific examples because they don't know how to do it. The textbooks typically endorse Darwinism in general terms in the introductory chapter and thereafter ignore it. Most molecular biologists accept Darwinism uncritically because they are scientific materialists and have no alternative, but the Darwinian mechanism plays no role in their science.

Science or Philosophy?

Most readers of this book probably don't feel qualified to judge scientific disputes. For that matter, Richard Dawkins himself is a zoologist and not a biochemist, and he told me himself that he doesn't feel qualified to debate Behe's scientific claims. What you and I and Dawkins can judge is whether Behe is right about the state of knowledge among molecular biologists, as reflected in the scientific literature. As scientists learn about the complexity of molecular mechanisms, do they find it possible to explain their origin by specific evolutionary pathways through functional intermediate stages? Or do they continue to believe in the existence of those pathways merely because their materialist philosophy allows no alternative?

An answer to that question may be found in the initial reactions of prominent scientists to Behe's book. Molecular biologist James Shapiro of the University of Chicago agreed with Behe that the Darwinian theory cannot explain molecular complexity; he wrote in *National Review,*

> There are no detailed Darwinian accounts for the evolution of any fundamental biochemical or cellular system, only a variety of wishful speculations. It is remarkable that Darwinism is accepted as a satisfactory explanation for such a vast subject—evolution— with so little rigorous examination of how well its basic theses work in illuminating specific instances of biological adaptation or diversity.

That sounds like a ringing endorsement of Behe's scientific claims, but Shapiro nonetheless blasted Behe for arguing that those unexplained biochemical systems might be designed. Raising that possibility was "fighting the battles of the past rather than seeing the vision of the future." That's another illustration of how strong the hold of materialist philosophy is on the minds of contemporary biologists. If Behe's science is accurate, why should the vision of the future exclude design?

Shapiro then proceeded from philosophical prejudice to a form of confusion we have seen before. What Behe failed to recognize, he wrote, was that we now have experience with computers. "Having exemplars of physical objects endowed with computational and decision-making capabilities shows that there is nothing mystical, religious, or supernatural about discussing the potential for similarly intelligent action by living organisms."

In a sense, that's perfectly correct. It's also another instance of Berra's Blunder. Those computers are intelligently designed. Unassisted matter never made a computer, nor did naturalistic evolution.

Although James Shapiro was confused about the concept of design, he did take the high road by considering Behe's arguments fairly. Another prominent University of Chicago biologist, Jerry Coyne, writing in the prestigious British journal *Nature*, took the low road of appealing to prejudice. Coyne began and ended his review with attacks on biblical fundamentalists, trying mightily to leave the impression that what is at issue is preserving the independence of science from religious control. Behe is a Roman Catholic who has no religious objection to Darwinian evolution; his argument is simply that the Darwinian mechanism has no scientific merit in molecular biology. Neither Shapiro nor Coyne contradicted Behe on any scientific point. Their objections were entirely philosophical, or based on a failure to comprehend the concept of design.

These reactions illustrate the thinking problem that I described in the preceding chapter. There are two definitions of science at work in the scientific culture, and a concealed contradiction between them is beginning to come out into public view. On the one hand, science is dedicated to empirical evidence and to following that evidence wherever it leads. That is why science had to be free of the Bible, because the Bible was seen to constrain the possibilities scientists were allowed to consider.

On the other hand, science also means "applied materialist philosophy." Scientists who are materialists always look for strictly materialist explanations of every phenomenon, and they want to believe that such explanations always exist. This raises the question: What will the scientists do if the evidence starts to point *away* from materialism and *toward* the possibility that a Creator is necessary after all? Will they follow the evidence wherever it leads, or will they ignore the evidence because their philosophy does not allow it to exist?

Most scientists won't discuss that possibility openly. One who has enough self-confidence to do it is the famous Harvard geneticist and Marxist Richard Lewontin, one of the most influential biologists in the world. Lewontin has written,

It is not that the methods and institutions of science somehow compel us to accept a material explanation of the phenomenal world, but, on the contrary, that we are forced by our *a priori* adherence to material causes to create an apparatus of investigation and a set of concepts that produce material explanations, no matter how counter-intuitive, no matter how mystifying to the uninitiated. Moreover, that materialism is absolute, for we cannot allow a Divine Foot in the door. The eminent Kant scholar Lewis Beck used to say that anyone who could believe in God could believe in anything. To appeal to an omnipotent deity is to allow that at any moment the regularities of nature may be ruptured, that miracles may happen.

In other words, evolution is not a fact, it's a philosophy. The materialism comes first (a priori), and the evidence is interpreted in light of that unchangeable philosophical commitment. If the evidence seems to go against the philosophy, so much the worse for the evidence. To a materialist, putting up with any amount of bad practice in science is better than to let that Divine Foot in the door!

Materialism and the Mind

The contradiction between materialism and reality arises frequently in biology, but it is most inescapable when we consider the human mind. Are our thoughts "nothing but" the products of chemical reactions in the brain, and did our thinking abilities originate for no reason other than their utility in allowing our DNA to reproduce itself? Even scientific materialists have a hard time believing *that*. For one thing, materialism applied to the mind undermines the validity of all reasoning, including one's own. If our theories are the products of chemical reactions, how can we know whether our theories are true? Perhaps Richard Dawkins believes in Darwinism only because he has a certain chemical in his brain, and his belief could be changed by somehow inserting a different chemical.

The absurdity to which this kind of reductionist thinking leads is marvelously illustrated by a story told by John Horgan, a writer for *Scientific American*. At a scientific conference, philosopher David Chalmers argued that a materialist science cannot explain human consciousness. His arguments were persuasive, and the scientists treated them respectfully, but they all wondered: What follows? If science cannot explain the mind, then what explanation can there possibly be?

Horgan explains, "Chalmers thought he had found a possible solution: scientists should assume that information is as essential a property of reality as matter and energy." As we saw earlier in this chapter, that is exactly what George C. Williams was led to believe in spite of his materialist philosophy. Horgan comments that Chalmers's matter-information dualism cannot be true, because science tells us that only the particles that make up matter and energy were present at the beginning. (That's an illustration of the fallacy of begging the question, because the assumption that matter necessarily comes before mind is what Chalmers was denying.) "Nevertheless," Horgan reports, "Chalmers's ideas struck a chord among his audience. They thronged around him after his speech, telling him how much they had enjoyed his message."

One listener was displeased, however. This was Kristof Koch, a dedicated materialist who collaborates with Francis Crick on brain research. Horgan went on to tell how the materialist dealt with the mystic:

That night Koch . . . tracked Chalmers down at a cocktail party for the conferees and chastised him for his speech. It is precisely because philosophical approaches to consciousness have all failed that scientists must focus on the brain, Koch declared in his rapid-fire German-accented voice, as rubberneckers gathered. Chalmers's information-based theory of consciousness, Koch continued, like all philosophical ideas, was untestable and therefore useless. "Why don't you just say that when you have a brain the Holy Ghost comes down and makes you conscious!" Koch exclaimed. Such a theory was unnecessarily complicated, Chalmers responded drily, and it would not accord with his own subjective experience. "But how do I know that your subjective experience is the same as mine?" Koch sputtered. "How do I even know you're conscious?"

Koch went on to admit to Horgan that he actually agreed with Chalmers that science cannot solve the consciousness problem. What this means to him is that consciousness must be meaningless or illusory. "How do I even know you're conscious?" he repeated to Horgan.

Indeed. For that matter, maybe Koch himself is permanently asleep and is just dreaming that he is conscious. Perhaps his thoughts are just illusions that his DNA has programmed into his brain to encourage him to make more DNA. A true-believing materialist will embrace even madness if the only alternative is to give up materialism.

So far we have seen that there is ample reason to believe that Darwinism is sustained not by an impartial interpretation of the evidence but by dogmatic adherence to a philosophy even in the teeth of the evidence. But can anything be done about this situation? The scientific establishment has immense power, particularly when it is supported by the media and the government. Critics can't get a fair hearing as long as Microphone Man filters everything they say through the *Inherit the Wind* stereotype. We need to think about strategy, and that's our next subject.

The Wedge

A Strategy for Truth

*T*he newspapers and radio shows were full of comment after Pope John Paul II sent a message to a meeting of the Papal Academy of Sciences in 1996 that seemed to endorse evolution. As usual, journalists interpreted this event strictly according to the *Inherit the Wind* stereotype.

For a few days the press portrayed the pope as a relatively enlightened religious leader who was willing to adjust his faith to the scientific facts, in contrast to obstinate Protestant fundamentalists who continue to fight for a literal interpretation of the first chapters of Genesis. The stories took for granted that a reasonably broad-minded religion can coexist peacefully with evolutionary science. Reporters forgot overnight that Richard Dawkins had just been touring the United States, carrying the more accurate message that Darwinism denies absolutely that a Creator is responsible for our existence.

In fact the pope's message was not quite as advertised. The

Roman Catholic Church long ago approved evolution as a scientific hypothesis worthy of investigation. On the other hand, the Catholic Church has consistently opposed materialism, as any genuinely Christian church must. John Paul II's statement continued this tradition. Although he remarked that the general principle of evolution has been supported by various independent lines of evidence, he also said that it seems there are several theories of evolution rather than only one, because there is substantial disagreement over both the mechanism and the philosophy.

Far from endorsing the materialist understanding of evolution that dominates contemporary science, the pope pronounced that "theories of evolution which, in accordance with the philosophies inspiring them, consider the spirit as emerging from the forces of living matter or as a mere epiphenomenon of this matter, are incompatible with the truth about man." As usual, Microphone Man kept the pope's real position from reaching the public by reporting only the sound byte "Pope Endorses Evolution."

To be fair to the reporters, the pope invited that kind of media treatment. He chose to state his crucial reservations in cautious language in the middle of a document whose central theme was conciliatory toward evolutionary scientists—most of whom do not acknowledge that there *is* any difference between science and materialism. If he had wanted to draw a line in the sand, the pope could have said bluntly in his first paragraph that evolution as understood by scientists like Richard Dawkins and Carl Sagan is based on materialist philosophy and hence incompatible with the truth about both God and humankind.

If he had done that, newspapers would probably have headlined the story "Pope Attacks Science," and the stories would have featured interviews with liberal professors at Catholic universities, expressing concern about the pope's increasing rigidity and lecturing him that "the Bible is not a science textbook." Accompanying editorials would have reminded their readers that the church persecuted Galileo centuries before and implied that vigilance will be necessary to make sure that the same thing doesn't happen again. Even a pope can be confined in a stereotype.

Should We Try to Accommodate?

It is understandable that the Vatican chose to emphasize the conciliatory side of the papal statement. Like most Christian leaders, the pope doesn't want to spend his time and energies arguing with scientists about their theories; he wants to give science its due and move on to the spiritual matters that primarily concern him. If currently accepted theories are true, they will prevail, and if they are not true they will eventually be discarded. So it may seem prudent to make peace with "evolution," provided that scientists are willing to interpret the theory in a way that doesn't rule out fundamental Christian doctrines like the resurrection or the creation of humankind in the image of God.

The trouble with the conciliatory strategy, however, is that it papers over a fundamental difference in worldview that can't be compromised. Scientific materialists genuinely believe that materialism and science are inseparable, that the realm of objective reality belongs entirely to science and that belief in a supernatural Creator is a holdover from the past that has no place in a rational mind. Religion is acceptable to materialists only as long as it stays in the realm of the imagination and makes no independent claims about objective reality. Creation must be a human way of thinking about what evolution has accomplished, and the resurrection must be an event that occurred only in the minds of the disciples. Many liberal Christian leaders have surrendered on those terms and have even become proponents of naturalism. Fortunately, the pope knows better than to do that.

In their own way, all Christian parents, teachers and students have to deal with the same problem the pope faced. The culture tells us that we have two alternatives. We can accept "evolution" as the scientists understand the term, which means that we implicitly accept naturalism and materialism (even if we pretend otherwise). Alternatively, we can reject evolution—in which case Microphone Man will stereotype us as premodern fundamentalists who insist on every detail of Genesis regardless of the evidence. Should we fight, or should we accommodate on the best terms we can get from the materialists?

If we choose to accommodate, we can take some advantage of the statements of science organizations, which do say (although not very convincingly) that science does not deal with religious questions or deny the existence of God. These halfhearted disclaimers create some wiggle room, especially considering that the more limited definitions of *evolution* may be perfectly consistent with Christianity, including even fundamentalist Christianity. Evolution within the species is as much a biblical doctrine as a scientific one, for the Bible taught us (long before modern science) that all the different races of humans descend from a common human ancestor. Finch-beak variation in no way denies that only God can make a bird.

We can even pretend, as some teachers do, that "all the scientists are saying is that all living things are related and that a certain amount of natural variation occurs in nature." *We* may know that Darwinian evolution is actually saturated with materialistic philosophy, but why not cooperate with the Darwinists by shrouding this fact in ambiguous words? Perhaps we can just say that we accept evolution as a scientific theory about *how* God created, and then drop the subject and go on to more pleasant matters. As a senator famously said during the unwinnable Vietnam War, why don't we just declare victory and go home?

Why Accommodation Doesn't Work
One answer to that question is that a shallow reconciliation of science and religion leaves our young people open to materialist indoctrination when they go away to college and learn there what "evolution" really means. Many readers of this book will

have seen the videotape of the debate I had with Cornell professor William Provine at Stanford University. Provine has told me that his father was a Christian minister who tried to reconcile Christianity and evolution through "process theology," the process being that God is evolving along with the world. When Provine studied evolution in graduate school and learned that it is a strictly materialist process that has no purpose or goal, he discarded the religious baggage and has been a dedicated atheist ever since.

The same has happened to many others, although usually less dramatically. As students grow more and more accustomed to assuming materialism and naturalism in their academic work, the concept of creation by God gradually tends to become less real to them. It is not so much that any single finding undermines their faith; rather, the day-to-day practice of thinking in naturalistic terms about academic subjects makes it awkward to think differently when it comes to religion. Young intellectuals may insist for years that they are still believers, but then one day they wake up to realize that their belief has been emptied of its content, and they either throw away the empty shell or fill it with something else. That is why every mainstream Christian institution is beset from within these days by people who want the church to turn away from the old business of sin and salvation and devote its energies to whatever social causes are currently fashionable in the secular world.

The reason shallow reconciliation doesn't work is that the specific conclusions of evolutionary science are only part of the problem, and the lesser part. God could work through evolution, or natural selection, and in limited respects he does. The greater problem is that modernist science protects its grand theory of evolution by starting with the basic assumption that God is out of the picture and by sticking to that assumption through every discouragement. When people are taught for years on end that good thinking is naturalistic thinking, and that bringing God into the picture only leads to confusion and error, they have to be pretty dense not to get the point that God must be an illusion. This doesn't necessarily mean that they become atheists, but they are likely to think about God in a naturalistic way, as an idea in the human mind rather than as a reality that nobody can afford to ignore.

Naturalism and Truth

Naturalistic thinking is as prevalent in the nonscientific departments of the university, including departments of religious studies, as it is in the sciences. It is as rare for a history professor to assert in professional circles that the resurrection might really have happened as it is for a biology professor to advocate intelligent design. Many literature professors have discarded the rationalism of science in favor of relativistic philosophies like deconstruction. They have retained, all the more desperately, the naturalism that frees them from having to worry about what God might think about their abandonment of truth. In fact, many of them invoke Darwinism to challenge the idea that there is such a thing as objective truth or an objective difference between right and wrong. According to the very influential philosopher Richard Rorty, "The idea that one species of organism is, unlike all the others, oriented not just toward its own increased prosperity but toward Truth, is as un-Darwinian as the idea that every human being has a built-in moral compass—a conscience that swings free of both social history and individual luck."

Rorty is absolutely right. Truth (with a capital *T*) is truth as God knows it. When God is no longer in the picture there can be no Truth, only conflicting human opinions. (There also can be no sin, and consciousness of sin is that built-in moral compass Rorty rejects as illusory.) We can know something about what is useful for getting whatever we happen to want, but false beliefs have often been extremely useful. In fact, modernists frequently cite belief in God as a prime example of a falsehood that has been useful for achieving social unity or comforting people against the fear of death.

The dream of modernists was that science would be an adequate substitute for Truth. This is the case, however, only with matters like the ordinary laws of physics (apples fall down, not up) which are subject to direct experimental testing. Very few really interesting propositions (like Darwinian macroevolution, for example) can be tested so directly and conclusively. With respect to these more elusive matters, scientific theories rely on elaborate reasoning and sophisticated interpretation and rest on assumptions that are difficult or impossible to prove.

At times science can even seem to make a mockery of reality. The most impressive scientific theory of contemporary times is quantum mechanics, which in some interpretations says that the exact location of a particle at a given time depends on whether somebody happens to be looking. Quantum mechanics unquestionably *works*, but whether and in what sense it can be said to be *true* has just about everyone baffled.

The result of such commonsense-defying theories has been to encourage speculation that the observer makes the world and even to foster the growth of intellectual movements that consider science itself to be only one way of interpreting the world. That speculation is the basis of what is called "postmodernism," which has become a formidable movement in the humanities. Modernists believe in a universal rationality founded on science; postmodernists believe in a multitude of different rationalities and consider science to be only one way of interpreting the world. In other words, modernists are rationalists; postmodernists are relativists.

Relativism is particularly hard to avoid in the realm of value, because one of the basic modernist assumptions is that "ought" cannot be derived from "is." Science may be able to tell us exactly how things happen, but it cannot tell us whether anything is bad or good, beautiful or ugly. Only humans (or God) can make moral or artistic judgments, and these judgments cannot be derived directly from mere facts. History may be able to tell us that most societies have condemned prostitution or homosexual behavior, but this fact cannot prove that such practices are "wrong" for us. After all, some of those same societies practiced barbaric cruelties and condoned slavery. Why shouldn't we toss tradition overboard and base our ethical and artistic standards on our own desires?

It is no good for parents to try to protect their children from the influence of thinkers like Carl Sagan or Richard Dawkins or Richard Rorty. The prominent modernist and postmodernist thinkers embody philosophical currents that permeate academia and the media at every level—in television series like *Star Trek*, for example. Even Christian college and seminary professors are bound to be influenced by the spirit of the times. To be successful in academic life is to be current with the fashionable thought from the most prestigious universities, and teachers can hardly help absorbing the ways of thinking that they themselves have been taught.

Taking a Stand

Protecting young people works only if they can be kept forever uninformed or unthinking, and that is a losing strategy in the long run. For that matter, it would be an unworthy strategy even if it were more successful. Jesus did not tell his disciples to form a protected community where they could shut out corrosive philosophies. He told them to "go and make disciples of all nations."

A faith that has to be protected behind walls is like a house built on sand. When the protection ceases, the faith collapses. Faith is confirmed by testing and validated by struggle in a world that gives a multitude of reasons for doubt. Instead of hiding our light under a bushel to protect it from the darkness, today we need to be more like the biblical men of Issachar, "who understood the times and knew what Israel should do."

If we understand our own times, we will know that we should affirm the reality of God by challenging the domination of materialism and naturalism in the world of the mind. With the assistance of many friends I have developed a strategy for doing this, and a major purpose of this book is to interest young people, and persons with influence over young people, in preparing themselves to take part in the great adventure we have begun.

Building the Wedge

We call our strategy "the wedge." A log is a seeming solid object, but a wedge can eventually split it by penetrating a crack and gradually widening the split. In this case the ideology of scientific materialism is the apparently solid log. The widening crack is the important but seldom-recognized difference between the facts revealed by scientific investigation and the materialist philosophy that dominates the scientific culture. What happens when the facts cast doubt on the philosophy? Will scientists and philosophers allow materialism to be questioned, or will they rely on Microphone Man to suppress the facts and protect the philosophy?

My own books (including this one) represent the sharp edge of the wedge. I had two goals in writing those books and in pursuing the program of public speaking that followed their publication. First, I wanted to make it possible to question naturalistic assumptions in the secular academic community. Second, I wanted to redefine what is at issue in the creation-evolution controversy so that Christians, and other believers in God, could find common ground in the most fundamental issue—the reality of God as our true Creator.

Protestants will disagree on various issues among themselves, Catholics will disagree with Protestants, and observant Jews will disagree with Christians. What all these should agree on is that God—not some purposeless material process—is our true Creator. Given that we inhabit a culture whose intellectual leaders deny this fundamental fact, we should unite our energies to affirm the reality of God. After we have had that positive experience of unity and affirmation, we may be able to talk about the remaining points of disagreement with renewed goodwill. This is the program I call *theistic realism.*

Michael Behe's book *Darwin's Black Box,* which I described briefly in chapter five, represents the first broadening of the initial crack in the scientific materialism "log." I first became acquainted with Behe when he wrote a letter to the editor of the journal *Science,* which had published a dismissive news article about me. I was naturally pleased to receive support from a reputable biochemist, and even more pleased that the letter was very well written. Subsequently, friends who were interested in promoting my ideas arranged an academic conference at Southern Methodist University, to which they invited ten scientists and philosophers, including Behe, to discuss the relationship between evolutionary science and philosophical naturalism.

Behe presented a paper on proteins at the conference and formed the idea of writing a book to demonstrate that biologists who study cells and molecular systems constantly see examples of irreducibly complex systems that cannot have formed by Darwinian evolution. A major New York trade publisher (Free Press) brought out Behe's book, indicating that our small movement was breaking out of the Christian ghetto and into the cultural mainstream.

The wedge is continuing to broaden. With the assistance of some generous donors and the staff of Christian Leadership Ministries, we put on a major conference on "Mere Creation" at Biola University in November 1996. Approximately two hundred persons attended, including scientists, philosophers and potential academic and financial supporters. Most were Christians, but the only requirement for attenders was a willingness and ability to contribute to the theme of the conference, which was that "the first step for a twenty-first-century science of origins is to separate materialist philosophy from empirical science." Sixteen persons gave papers, and of course there was extensive discussion about the next steps on our intellectual agenda.

What is next on the agenda? Scientifically, there is the question of how far the reconsideration of Darwinism is going to take us. We know that the Darwinian mechanism doesn't work and that complex biological systems never were put together by the accumulation of random mutations through natural selection. This is not a mere gap in a theory that is sound in other respects. It isn't just that the Darwinists have failed to provide a complete explanation; they've failed even to understand what needs to be explained. Their theory assumes that *variation* is all they need to explain and that the accumulation of small variations over immense amounts of time can produce complex organisms from simple beginnings. That is why they think that finch-beak variation illustrates the process that created birds in the first place.

Once the problems of informational content and irreducible complexity are out on the table in plain view, well-informed people are going to be amazed that scientists took so long to see that random mutation is not an information creator and that the Darwinian mechanism is therefore irrelevant to the real problem of biological creation. A few scientific materialists are aware of this and hope to rescue the situation by discovering new information-creating laws of physics and chemistry. Good luck to them, but the prospects are about as promising as the prospects of finding new laws of ink and paper that can create Shakespeare's plays.

Granted that the materialist mechanism has to be discarded, what does this imply for what scientists call the "fact of evolution," the concept that all organisms share a common ancestor? Universal common ancestry is as much a product of materialist philosophy as is the mutation/selection mechanism. Consider the proposition that a single ancestral bacterium gave birth to distant descendants as diverse as trees, insects and birds. If materialism is true, then universal common ancestry virtually has to be true also. The only materialist alternative is that life arose from nonliving chemicals many separate times, and this seems not only improbable but inconsistent with the observable fact that all living organisms share a common biochemistry. Life seems to have arisen from a single source, and if materialism is true, that source must have been a material ancestor.

Put aside the materialism, however, and the common ancestry thesis is as dubious as the Darwinian mechanism. There is no known process by which a bacterial species can evolve the immense complexity of plants and animals—in fact there is only a beginning of an understanding of what that complexity involves. There is no fossil history of single-celled organisms changing step by step into complex plants and animals. On the contrary, the major groups of animals all appear suddenly in the rocks of the Cambrian era—and no new groups appear thereafter. (High-school textbooks either fail to mention this fundamental fact of the fossil record or refer to it so obliquely that students don't see the implications.) The fossil problems are only the beginning, however, because evidence from embryology and genetics is adding to the difficulties.

This is not the place to develop the scientific ideas further; my purpose here is just to give a hint of the excitement that animated the scientists and philosophers who attended the Mere Creation Conference. The British scientific materialist J. B. Haldane wrote years ago, "My own suspicion is that the universe is not only queerer than we suppose, but queerer than we *can* suppose." For some obscure reason, Darwinists like to quote that statement, although Darwinism asserts that the realm of life is not queerer than we can suppose but at bottom very simple and commonsensical. All it takes to make a world of living things, according to the theory, is variation, natural selection, changing environments and long periods of time. But that is nineteenth-century science, and it won't survive the opening of *Darwin's Black Box.* When biology finally has its quantum revolution, our view of life and its origin will change profoundly.

Even more exciting than the scientific part of the theistic realism agenda, at least to me, is the new understanding of rationality that it promises. (This is the subject of my book *Reason in the Balance.*) Materialism tells us, incredibly, that the universe can be rational only if it is the product of impersonal laws, and not if it is the creation of a supreme mind. Materialists tend to think the only alternative to materialism is some form of primitive superstition, where science would be impossible because all events would be produced by the whimsy of capricious gods. This is nonsense, of course. Intelligent design does not mean unintelligent chaos. Computers and space rockets are designed, but they work according to lawlike principles.

The real objection scientific materialists have to design is that the Designer would be something outside of science and hence not subject to human control. The attraction of a materialistic universe is that it feeds the imperialism of science by seeming to promise that everything can in principle be understood (and controlled) by science. There is an immense price to be paid for this illusion that we can have a "theory of everything," however. There can be no science of value, or of beauty, or of goodness. The whole realm of value is left to the subjective imagination, with destructive consequences that we can see all around us. Eventually materialist philosophy undermines the reliability of the mind itself—and hence even the basis for science. The true foundation of rationality is not found in particles and impersonal laws but in the mind of the Creator who formed us in his image.

Probably many readers of this book feel that pursuing the intellectual program of theistic realism, or even completely understanding it, is beyond them. No matter if it is; nearly everyone knows some young person who has the necessary gifts. I find as time goes by that my greatest satisfaction comes not from the work I can do myself but from the accomplishments of younger people to whom I have given encouragement and for whom I have opened doors. If you know a gifted young person, help him or her to see the vision. Those who are called to it won't need any further encouragement. Once they have seen their calling, you had better step out of the way because you won't be able to stop them even if you try.

Modernism

The Established Religion of the West

*T*hree significant events in recent American history mark the culmination of a fundamental change that occurred gradually in U.S. society—and is evident in other Western societies as well—over the course of the twentieth century.

The Darwin Centennial Celebration
The first of these events was the great Darwin Centennial celebration of 1959, commemorating the publication of *On the Origin of Species* one hundred years earlier. The celebration was held at the University of Chicago, which had been the site of two other scientific milestones of the mid-twentieth century. One of these was the first self-sustaining atomic chain reaction, at a primitive reactor underneath the university's abandoned football stadium in 1942. The second was a famous experiment by chemist Stanley Miller in 1953, which had produced amino acids by sending electrical current through a mixture of gases. Although the Miller experiment proved to lead only to a dead end, at the time

it gave scientists confidence that they would soon discover how life evolved on the early earth from nonliving chemicals.

The participants in the Darwin Centennial were understandably in a triumphal mood. The prestige of science was never higher. Polio had been conquered by a vaccine; atomic power seemed to promise abundant, cheap energy; space travel loomed in the near future. Besides these technological achievements, science had seemingly established that a purposeless process of evolution was our true creator and hence had dethroned the God of the Bible. The religious implications of this intellectual revolution were frankly emphasized by the most prominent speaker at the centennial, the British biologist, philosopher and world statesman Sir Julian Huxley.

Julian Huxley was the grandson of Thomas Henry Huxley, who was known as "Darwin's bulldog" because he was the most important early champion of Darwin's theory. T. H. Huxley had also invented the word *agnostic* to describe his own religious views. Julian Huxley, a zoologist, was one of the scientific founders of the neo-Darwinian synthesis, the modern version of Darwin's theory. He was also the promoter of a naturalistic religion called evolutionary humanism, and the founding secretary general of UNESCO, the United Nations Educational, Scientific and Cultural Organization. In short, Julian Huxley was one of the most influential intellectuals of the mid-twentieth century,[1] and 1959 was the high-water mark of his influence. Here are some excerpts from Huxley's remarks at the centennial:

> Future historians will perhaps take this Centennial Week as epitomizing an important critical period in the history of this earth of ours—the period when the process of evolution, in the person of inquiring man, began to be truly conscious of itself. . . . This is one of the first public occasions on which it has been frankly faced that all aspects of reality are subject to evolution, from atoms and

---

[1]He was probably also the model for Jules, the figurehead leader of the sinister N.I.C.E. in C. S. Lewis's classic futurist novel *That Hideous Strength* (New York: Macmillan, 1946).

stars to fish and flowers, from fish and flowers to human societies and values—indeed, that all reality is a single process of evolution.

In 1859, Darwin opened the passage leading to a new psychosocial level, with a new pattern of ideological organization—an evolution-centered organization of thought and belief.

In the evolutionary pattern of thought there is no longer either need or room for the supernatural. The earth was not created, it evolved. So did all the animals and plants that inhabit it, including our human selves, mind and soul as well as brain and body. So did religion.

Evolutionary man can no longer take refuge from his loneliness in the arms of a divinized father figure whom he has himself created, nor escape from the responsibility of making decisions by sheltering under the umbrella of Divine Authority, nor absolve himself from the hard task of meeting his present problems and planning his future by relying on the will of an omniscient, but unfortunately inscrutable, Providence.

Finally, the evolutionary vision is enabling us to discern, however incompletely, the lineaments of the new religion that we can be sure will arise to serve the needs of the coming era.

In short, the triumph of Darwinism implied the death of God and set the stage for replacing biblical religion with a new faith based on evolutionary naturalism. That new faith would become the basis not just of science but also of government, law and morality. It would be the established religious philosophy of modernity.

Inherit the Wind

The 1960 film version of *Inherit the Wind* was essentially the artistic equivalent to the 1959 Darwin Centennial. It portrayed the triumph of Darwinism as a Hollywood-style political liberal of the period would have seen it. The forces of freedom and enlightenment defeated the forces of ignorance, represented by Christian fundamentalism, and thus allowed the young lovers to escape to a better world.

Because I have already devoted a chapter to the play and film, I will say little more about it here except to remind you of the

importance of the final scene. At the very end of the film the
wise defense lawyer, played by Spencer Tracy, weighs the Bible
and *On the Origin of Species* in his hands, shrugs and then puts the
two books together in his briefcase. The implied message is that
the two are equivalent and compatible. The Book of Nature and
the Word of God are in agreement, provided the latter is interpreted in the light provided by the former. The closing gesture
assures the audience that Darwinian naturalism does not aim to
abolish Christianity but to liberalize it so that it is compatible
with a properly scientific understanding of our origins. Fundamentalist resistance to evolution is thus shown to be not only
unintelligent and futile but also unnecessary.

The liberalized Christianity implied by this final scene in
Inherit the Wind has been far more effective in legitimating
evolutionary naturalism than the explicit atheism of Richard
Dawkins or Julian Huxley's proposed new religion of evolutionary humanism. Why repudiate Christianity explicitly when its
rituals and language can be taken over and given a naturalistic
meaning? The death of God does not require the end of religion
or even the end of the traditional Christian denominations. On
the contrary, the new religion Huxley foresaw was already se-
curely established within mainline Christian denominations.
Liberal ministers and theologians try to save Christianity by
"demythologizing" it—removing or downplaying those supernatural elements that are so embarrassing to modernists.

It is fairly easy to do this without openly denying key doctrines
like the resurrection, because modernists tend to interpret religious statements as something like poetry. When a poet writes
about miracles, scientific naturalists will take no offense, because
they know that poetry is meant to convey the feelings of the poet
rather than the facts of nature. Likewise, it is possible for a
minister or seminary professor to speak with great feeling about
the resurrection while signalling to the philosophically sophisticated that the event occurred only in the minds of the disciples.

Politically astute scientific naturalists feel no hostility toward those religious leaders who implicitly accept the key naturalistic doctrine that supernatural powers do not actually affect the course of nature. In fact, many scientific leaders disapprove of aggressive atheists like Richard Dawkins, who seem to be asking for trouble by picking fights with religious people who want only to surrender with dignity.

Besides, debating the truth or falsity of religious claims takes those claims more seriously than they deserve. To say that a statement is false is to concede that it could conceivably be true. This can be dangerous. Focusing the mind of an unbeliever on the question whether Christ's claims are true has often had unanticipated consequences. The most sophisticated naturalists realize that it is better just to say that statements about God are "religious" and hence incapable of being more than expressions of subjective feeling. It would be pretty ridiculous, after all, to make a big deal out of proving that Zeus and Apollo do not really exist.

The School Prayer Decision
The third defining event of the mid-twentieth century was the Supreme Court's 1962 decision in *Engel* v. *Vitale,* which banned officially prescribed prayers from the public schools as an "establishment of religion." I am not concerned here with whether officially promulgated school prayers are a good thing, but with what the context of the decision tells us about changing attitudes toward God.

The prayer in question came not from the Bible Belt but from New York, a state with a large and influential Jewish population and a liberal tradition. Far from being oppressive in purpose, it represented a well-intentioned effort by public school officials to bring Jews and Christians together on the basis of the theism that was thought to unite them. The approved prayer read simply, "Almighty God, we acknowledge our dependence upon Thee, and we beg Thy blessings upon us, our parents, our teachers, and our Country." The Pledge of Allegiance had just recently been amended to affirm that Americans are "one nation, under God," and so educators had good reason to think that a simple affirmation of our dependence on a common Creator would be uncontroversial.

They were mistaken. By 1962 "God" in intellectual circles was a discredited concept associated not with education and social unity but sectarian conflict and superstition. Although America had been remarkably free of religious strife and had welcomed millions of Catholic and Jewish immigrants to what had once been an overwhelmingly Protestant country, the centuries of religious wars and persecutions in Europe had given religion a bad name among intellectuals. Bad historical memories were reinforced by modernist philosophy, according to which God is the subjective creation of human culture. This implies that each religious or cultural group effectively worships a different deity of its own creation.

A God who is not the same for everyone cannot unite diverse peoples. Unity must be achieved through a common way of thinking based on scientific reasoning—which *is* the same for everyone. For American public schools, such a common rationalism was already available, having been prescribed by the immensely influential agnostic philosopher John Dewey. (Like Julian Huxley, Dewey consciously saw himself as promulgating a new religion, one that would be established as the basis of government and public education.) Whether students recited a prayer or not, public education aimed to teach them to rely on human intelligence and scientific methodology rather than take refuge in the arms of a divinized father figure who exists only in the human imagination.

The Supreme Court's school prayer decision thus merely ratified a transformation that had already occurred in the minds of the most influential educators. I am not inclined to protest the decision itself, because the prayer could have been a meaningless ritual even if the Supreme Court had approved it. It would have been something like the unenthusiastic required weekly singing of "God Save the Queen" I witnessed when teaching at an East African school shortly before the end of British colonial rule. Just as the African students made it painfully evident that they didn't care whether God saved the queen or not, many New York students would have found a way to express their disdain for a religious ritual that the school system itself did not take seriously.

If educators really believed that we are dependent upon God, they would spend time on the subject in the classroom instead of relegating it to a perfunctory ritual. Modernist educators agree with religious people that it is important for students to know who or what created them; that is why they insist on the teaching of evolution as fact.

A New Declaration of Independence

It would be roughly accurate to say that the 1960s marked the second American Declaration of Independence, our declaration of independence from God. One might expect far-reaching moral and legal consequences to follow from such a declaration, and so they did. Before the mid-twentieth century, most Americans assumed that the law was based on a set of underlying moral principles that came ultimately from the Bible. Protestants, Catholics and Jews differed on theological points, but on moral questions they were in broad agreement. For example, concepts about the sanctity of marriage which today are very much in doubt were taken for granted. Divorce was discouraged, both by law and by social pressure, and educators up to and through the university level did what they could to prevent premarital sex. The underlying moral code rarely had to be defended because it was rarely challenged. There was plenty of hypocrisy of course, and some elites (like movie stars) lived by different standards, but the rules were much as they had been a century earlier.

The change took hold in the late 1960s, as the new religious assumptions that had been gradually gaining ground began to have practical effects. When God's existence is no longer a *fact* but a subjective *belief* (and a highly controversial belief at that), God's moral authority disappears. It is no coincidence, therefore, that a drastic change in the nature of marriage immediately followed the change in the ruling philosophy. Both the legal restrictions on divorce and the social stigma evaporated practically overnight. Marriage ceased being a sacred covenant involving God and the community as well as husband and wife. It became an ordinary contract that could be ended by either party practically at will. What used to be called illegitimacy became respectable as single parenting, and the traditional two-parent household even began to seem ridiculous, a pathetic attempt to emulate an *Ozzie and Harriet* dream family that had never existed in reality.

With the divorce revolution came the sexual revolution, as the death of God and the availability of contraceptives seemed to make chastity obsolete. Hard on the heels of the sexual revolution came the feminist revolution, with a radical wing that explicitly rejected the traditional family model that had previously been regarded as the backbone of society. Feminism demanded an unrestricted right to abortion, which the Supreme Court duly read into the Constitution and imposed on a reluctant nation. Homosexual liberation came next, and homosexual activists quickly gained "victim" status and consequent support for their cause from the media. The Supreme Court again fell compliantly in line with the cultural trend, managing to find in the Constitution a principle that laws based on "animosity" toward homosexuality are unconstitutional. The moral and legal reversal was unstoppable once the crucial change in the established religious philosophy had been made.

The point is not that people are less moral today than they were previously, but that their morality took a different direction when its foundation shifted. Modernists can be as firm in their moral convictions, and as legalistic in enforcing them, as were the fundamentalists who ruled the fictional town of Hillsboro in *Inherit the Wind.* For those who are on the receiving end of it, "political correctness" is just as coercive as traditional religion, and just as capable of stifling free thinking. At Harvard as at Hillsboro, there are truths that only a very courageous teacher would dare to say in a classroom.

Modernists have also proved themselves willing to erect legal barriers to ensure that only the established view of religion is taught in the public schools. If at one time it was illegal in a few states to teach evolution, now it is considered unconstitutional in all states to teach or advocate creation as an alternative to evolution. As we have seen, in 1987 a majority of the Supreme Court held that it is unconstitutional for a state to provide for the presentation of a creationist alternative to evolution in the schools, because to do so would "advance the religious viewpoint that a supernatural being created humankind." The question whether that viewpoint might be true did not arise, because the majority assumed the modernist position that religious beliefs are about feelings, not facts. Justice Scalia argued in dissent that the people are entitled "to have whatever evidence there may be against evolution presented in their schools." His position baffled the modernists who dominate the legal culture. What evidence could there conceivably be against a scientific fact?

Politics Is Not the Answer
People who are dissatisfied with these developments frequently try to reverse them by becoming involved in partisan politics or issuing quixotic demands for the impeachment of Supreme Court justices. That kind of political activity has been spectacularly unsuccessful. Indeed, many of the path-breaking judicial opinions that social conservatives complain about were authored by justices appointed by such conservative presidents as Eisenhower, Nixon, Reagan and Bush. Political action may slow down the rate of change, but eventually the logic of the ruling philosophy will prevail.

At the moment, for example, a majority of Americans assume that marriage is by nature a heterosexual relationship and that a "marriage" of two men or two women is a contradiction in terms. That is why President Clinton reluctantly signed the 1996 Defense of Marriage Act, defining marriage for purposes of federal law as a union of a man with a woman. Opinion leaders in the intellectual world, however, probably including President Clinton himself, view this act as an exercise in bigotry, much like the laws that once prohibited interracial marriage. Exactly what gives a majority the right to enforce a particular religious viewpoint about marriage, when that viewpoint is constantly being called into question not only in secular institutions but even in mainstream churches? That's a tough question to answer, especially if you have to stick to modernist criteria.

The real question is whether the modernist criteria are right or whether we are in the grip of a misguided intellectual fashion that is leading us straight into unreality. Addressing that question is the job not of a mass political party but of an intellectual and spiritual movement.

I regard the idea of a Christian political party with a combination of horror and amusement, because Christian denominations are themselves so confused and internally divided. Naturalistic thinking is nearly as prevalent in the religious world as in the secular culture. I belong to the mainline Presbyterian (PC-USA) denomination myself, and we are having quite enough trouble trying to get our own denomination back on the right road without trying to govern the world in general.

Politics is not the answer, but that isn't a counsel of despair. On the contrary, this should be a time of excitement because it is a time of great opportunity. Christianity has always thrived on adversity. What it can't stand is worldly success and social respectability. The Christian philosophy that was overthrown in the 1960s was an easy target because it had become identified with American culture and with worldly ideas like human perfectibility and the inevitability of progress, which are actually profoundly un-Christian. The agnostics are not to be blamed for moving into the resulting vacuum; on the contrary, I credit them with helping to clear out some of the rot that has infested the timbers of the house of God. In an age in which people have learned to be distrustful of established institutions of all kinds, being kicked out of the establishment has its advantages.

Just about everywhere in the Christian world today, there is a combination of decay at the top and vitality at the bottom. (Thank God it isn't the reverse!) Denominational bureaucracies and seminaries are desperately in need of thoroughgoing renewal, while the pews and parachurch organizations are filling up with dedicated and talented people. The dedicated people have a chance to speak to a secular society that isn't as confident as it was in 1960 and to an intellectual community that is itself confused and divided over the unanticipated consequences of modernism. That's just the sort of challenge and opportunity they ought to welcome.

Despite decades of propaganda in the media and indoctrination in the schools, most Americans are skeptical of the philosophy of evolutionary naturalism and materialism. They are also well aware that this philosophy has not led to the era of rationality and social progress that was predicted. Even in the universities, where there is a separate culture war raging between scientific rationalists and postmodernists, there is a growing awareness that the ideas of 1960 are ripe for reconsideration. Western society will soon be ready to listen to a better idea. The question is whether we will have one to offer.

8

Stepping off the
Reservation

_____

*B*illy Graham began his career as an evangelist in the 1940s,
partnering in the early days with another gifted young man
named Charles Templeton. Templeton went on to study at
Princeton Theological Seminary, where he encountered the
"higher criticism" that American scholars had imported from
Germany. This naturalistic approach to biblical interpretation
assumed a scientific worldview that ruled out miracles and
viewed the Bible strictly as a product of human traditions. Modernist scholars sought to discover a "historical Jesus" who, unlike
the Jesus of the Gospels, worked no miracles and died like other
men.

Templeton began writing letters to Graham, urging him to
upgrade his amateurish theology by learning what modernist
scholars had discovered. Graham was almost overwhelmed by
Templeton's arguments, coming as they did with the full

weight of academic authority behind them. He came to a fork in the road. He could accept Templeton's challenge and devote years of study to the questions the modernist scholars were raising, or he could continue as a popular evangelist preaching the gospel. Either choice would change him for life. If he didn't immerse himself in modernist biblical criticism, he would never fully understand what he was rejecting. If he did pursue that kind of mind-bending study, he likely would not retain his unique gift to speak to the hearts of ordinary people.

After praying over the choice, Graham concluded, "I don't have the time, the inclination, or the set of mind to pursue [the intellectual questions]. I found that if I say 'The Bible says' and 'God says,' I get results. I have decided I'm not going to wrestle with these questions any longer."

Having made his decision, Billy Graham never looked back. He went on to become the twentieth century's greatest winner of souls for Christ and one of the world's most admired men. Charles Templeton left the Christian faith and became an agnostic. He is newsworthy today only because of his early association with Billy Graham.

Templeton charged Graham with having committed intellectual suicide, although he admitted that his friend would not have been so effective a preacher if he had allowed his message to be compromised by doubt. When Time magazine retold the Graham-Templeton episode for a 1993 cover story, it included a third character, who stands as an example of what Graham might have become if he had taken Templeton's advice. This was the modernist Episcopal bishop John Shelby Spong, who had delivered newspapers to the Graham family farm as a boy in North Carolina.

Spong is a classic example of a familiar figure in liberal Christian circles, the intellectual who rebelled as a young man against a fundamentalist upbringing and swung wildly to the opposite extreme. Having embraced modernist naturalism as the standard of truth,[1] Spong wants to "save" Christianity by purging it of its supernatural elements so that the gospel message of love and generosity can be credible to modernist minds. Spong commented to a *Time* writer, "I would never seek to solve the ethical problems of the 20th century by quoting a passage of Holy Scripture, and I read the Bible every day. I wouldn't invest a book that was written between 1000 B.C. and A.D. 150 with that kind of moral authority." That message never filled a stadium with sinners primed to walk up the aisle and accept Jesus.

When Charles Templeton said that Billy Graham had committed intellectual suicide, he meant that Graham had turned away from the true reality shown to us by our materialist science and modernist biblical criticism and had chosen instead to live in a world of illusion. In the world as modernists understand it, only matter existed at the beginning. Human beings did not fall from perfection into sin but evolved from savagery to civilization. Sin itself is an illusion, a guilt trip imposed by manipulative religious authorities. The way to meet humanity's needs is to provide enlightened social programs and therapies guided by scientific knowledge. According to modernism, this world is all there is, and the rational person aims to enjoy it and perhaps improve it a little before entering the oblivion of the grave.

Billy Graham's world is the opposite of all that. In the beginning was the Word, and nothing was created except by the Word. The fundamental fact about the human situation is that we are captured by sin, and we cannot escape from sin by our own efforts, however enlightened and humane those efforts may be. That is why the Word had to become flesh and dwell among us. There really is a heavenly Father and a risen Savior who can save fallen souls, and Billy Graham at the end of his earthly life can

---

[1]Spong's theology has been heavily influenced by his understanding that all historical statements must be judged by materialist criteria. For example, he explained in a newspaper interview that he rejects the doctrine that Jesus ascended to heaven because "Carl Sagan is a friend of mine. He said that if Jesus ascended literally and traveled at the speed of light, he hasn't yet gotten out of our galaxy" (from *Arizona Republic*, March 9, 1996, p. B5).

look forward to death because it means eternal life with that Savior. That is Billy Graham's world of sin and salvation, the world he refused to exchange for the world his friend was entering. If the main features of Graham's world are real, then Charles Templeton is the one who committed intellectual suicide.

Templeton, Spong and Graham all realized that the conflict between the naturalistic worldview and the Christian supernaturalistic worldview goes all the way down. It cannot be papered over by superficial compromises, such as Emilio's three mistakes. It cannot be mitigated by reading the Bible figuratively rather than literally. From a modernist perspective, biblical Christianity is just as wrong figuratively as it is literally. The story of salvation by the cross makes no sense against a background of evolutionary naturalism. The evolutionary story is a story of humanity's climb from animal beginnings to rationality, not a story of a fall from perfection. It is a story about recognizing gods as illusions, not a story about recognizing God as the ultimate reality we are always trying to escape. It is a story about learning to rely entirely on human intelligence, not a story of the helplessness of that intelligence in the face of the inescapable fact of sin.

There is no satisfactory way to bring two such fundamentally different stories together, although various bogus intellectual systems offer a superficial compromise to those who are willing to overlook a logical contradiction or two. A clear thinker simply has to go one way or another.

From that common understanding, Templeton, Spong and Graham made their separate choices. Templeton left the Christian faith altogether. Bishop Spong set out to transform Christianity into a creed more like the evolutionary humanism of Julian Huxley, dedicated to good works and enlightened social policy under the guidance of science. Billy Graham put aside the intellectual doubts and preached the gospel, going on to bring vast multitudes to Christ.

Billy Graham symbolizes the achievement of twentieth-cen-

tury Christianity and also its tragedy. Despite formidable obsta-
cles, the gospel faith held a place in humanity's heart, but it
virtually abandoned the mind to naturalism and materialism. Of
course many Christians did retain their faith while pursuing
academic careers, but they did so by finding some way to resist
or ignore the philosophical currents that dominated the academic world. Doubting his own intellectual powers, Graham
saved his faith and his effectiveness by stepping back from the
intellectual battle and engaging only the heart. For him, that was
surely the right choice. If he had tried to come to grips with the
modernist worldview, he might have only diluted the clarity of
his clear-cut presentation of the gospel.

Even the great British literary scholar C. S. Lewis, who certainly did not abandon the mind, spoke to Christians largely
through popular books and children's stories. Brilliant and
learned as he was, not even Lewis knew how to take on the
scientific elite during the high tide of materialism in the middle
of the twentieth century.

These great men, along with many others of lesser renown,
did the best they possibly could in the circumstances in which
God had placed them. It was not the first time in Christian history
that the faith did not appeal to intellectuals. In fact, the gospel
was every bit as contrary to the worldly wisdom of the first century
A.D. as it is to that of the twentieth century. As the apostle Paul
told the Christians at Corinth, "Not many of you were wise
according to worldly standards, . . . but God chose what is foolish
in the world to shame the wise" (1 Cor 1:26-27).

Eventually those who in Paul's time were counted foolish in
the world actually did shame the wise. It may happen again, and
a lot sooner than you might think. The world at the beginning
of the twenty-first century is not the same as the world Billy
Graham and C. S. Lewis faced in the middle of the twentieth
century. In their day modernist thought was everywhere trium-
phant and full of pride. Now materialist rationalism has just
about exhausted its potential.

Every history of the twentieth century lists three thinkers as preeminent in influence: Darwin, Marx and Freud. All three were regarded as "scientific" (and hence far more reliable than anything "religious") in their heyday. Yet Marx and Freud have fallen, and even their dwindling bands of followers no longer claim that their insights were based on any methodology remotely comparable to that of experimental science. I am convinced that Darwin is next on the block. His fall will be by far the mightiest of the three.

Evolutionary biology is a field whose cultural importance far outstrips its modest intellectual and scientific content. Its sacred trust is to preserve the central, indispensable part of the modernist creation story, which is the explanation of how such things as life, complex organ systems and human minds could exist without a Creator to design and make them. We might say that the point of Darwinism is to refute the otherwise compelling teaching of Romans 1:20, which is that God's eternal power and deity have always been evident from the things that were created. If Darwinism is in serious trouble—trouble that can't be fixed by a Band-Aid solution like a new variation on the mutation/selection mechanism—then the proud tower of modernism is resting on air.

No one in our day should find it hard to believe that a cultural tower built on a materialist foundation can look extremely powerful one day and yet collapse in ruins the next. We saw it happen when the Soviet Union broke apart because even its leaders lost faith in the ideology that had sustained it. The crimes of communism never discredited Soviet power. What destroyed that power was a loss of confidence, a loss of the assurance that some infallible "science" guaranteed that communism, regardless of its crimes and errors, would inevitably inherit the earth.

Darwinism in the West is in much the same condition as was Soviet Marxism in its last days. Its power and prestige rest not on any real scientific accomplishments but on the theory's role in upholding the ruling philosophy. Obscure scientists who go to a remote island to measure finch beaks can become the subjects of television documentaries and Pulitzer Prize-winning books, because the intellectual elite relies on finch-beak variation to convince the public that materialism is true. The biologists are at each other's throats in private, fighting over every detail in the Darwinist scientific program. The versions of "evolution" promulgated by Richard Dawkins and Stephen Jay Gould, for example, have hardly anything in common except their common adherence to philosophical materialism and their mutual dislike for supernatural creation. The full story of those conflicts never really reaches the masses, however. Microphone Man knows when to go back to the studio for soothing music and when to tell the listeners that "religious fundamentalists are attacking science again."

In short, evolutionary science has picked up the bad habits Richard Feynman warned scientists against and has thereby learned to fill an impressive balloon with hot air. To collapse the balloon, one only needs to make a tiny hole in its outer layer and let out some of the overconfidence that leads materialists to believe, as Marxists did, that history guarantees that their philosophy will overcome its problems and triumph in the end. Once that happens, I predict that the theory will collapse with astonishing swiftness.

The beginning of the end will come when Darwinists are forced to face this one simple question:

> *What should we do if empirical evidence*
> *and materialist philosophy*
> *are going in different directions?*

Of course the two are going in different directions, and much of the overelaborate baggage of Darwinism (punctuated equilibria, Berra's Blunder) exists only because it helps the Darwinists avoid seeing the fact that would otherwise be staring them in the face.

If you want to challenge Darwinian materialism, don't worry about anything else—just push this question and refuse to accept the usual evasions as answers. Tell every Darwinist you know that you won't be satisfied until Richard Dawkins and Stephen Jay Gould agree to take that question seriously, and to answer it in front of a critical audience that knows both the scientific background and the standard rhetorical moves.[2] The biologists have to tell us candidly whether they are asking us to believe in materialism because of what they know from studying the facts of biology or whether they are so devoted to the philosophy that they are willing to disregard evidence that doesn't fit it.

If the materialist domination of the intellectual world is seriously called into question, it will be possible for the next generation of Christians to enter the universities as participants in the search for truth, not as outsiders who have no choice but to submit to materialist rules. Instead of retreating from the public world of reason into the protected territory of faith, they will be pressing the questions that need to be pressed. Here are just a few of them: Why should the life of the mind exclude the possibility that a mind is behind our existence? Why should we assume that modern materialist philosophies are the wave of the future instead of a holdover from the nineteenth century? If information is something fundamentally different from matter, what is the ultimate source of the information? Will science be harmed if it gives up its ambition to explain everything, or has that ambition only harmed science by tempting scientists to resort to unsound methods? If materialism is not an adequate starting point for rationality, what alternatives are there?

Unfortunately, public education isn't doing much to prepare our young people for the twenty-first century. It's still under the domination of the old philosophy and isn't even teaching *that* very well. Parents increasingly realize that they have to take

---

[2]When someone claims to have magical powers, the claims must be tested before an audience of stage magicians, who know how the tricks of illusion are done. Scientists are notoriously easy to fool in such matters. When dealing with an ideology like Darwinism, the critical audience needs to include professors of rhetoric and legal scholars, who are skilled at spotting question-begging assumptions and similar tricks of logic.

primary responsibility for their children's education themselves, even if they use the public schools to do part of the job.

The main thing Christian parents and teachers can do is to teach young thinkers to understand the techniques of good thinking and help them tune up their baloney detectors so they aren't fooled by the stock answers the authorities give to the tough questions. When high-schoolers hear the word *evolution*, particularly in one of those public television science programs, the indicator screens on their baloney detectors should display "Snow Job Alert! Snow Job Alert!" If you know a good science teacher of the Bert Cates sort, tell him or her what you are doing. Maybe the teachers will want to learn more about the "snow job" themselves, so that they can teach science as a way of opening minds rather than as a process of memorizing the official story.

One problem you and they may have is figuring out how to avoid attracting the attention of so-called civil liberties lawyers, who actually specialize in confiscating baloney detectors. If you encounter lawyers like that, maybe my colleagues and I can help. My experience is that most such people do not act out of bad motives but because they have been mistaught. Perhaps we can set up educational forums to explain how we can open up the subject of evolution without opening the door to the Jeremiah Brown-style religious fanaticism that people rightly want to keep out of the schools.

It will be a big help in doing that if the people who want to challenge the official story are careful to do so in the right way. I'll put it figuratively: before you do anything else, be sure you audition for the right role in the next revival of *Inherit the Wind*. Bert Cates and Henry Drummond are going to win the case again, and Jeremiah Brown is going to lose. Matthew Harrison Brady is going to be shown up again as a dogmatist who isn't as smart as he thinks he is. What may change is that at the beginning of the twenty-first century, these characters will not be on the same side of the evolution controversy as they were in the 1920s. Even the Supreme Court may realize that

Hillsboro isn't the same place it used to be.

As always, we had better count the cost before we start to build the tower. If we are going to encourage baloney detecting, then we must be prepared to live with the consequences, *all the way down*. The scientific materialists thought they could encourage skepticism about everything else but keep their own core doctrines safe by identifying them with an infallible "science." I hope to show them that they were misguided and that critical thinking, once encouraged, can't be restricted.

It is the same with Christians. We shouldn't expect our more assertive young people to stop thinking for themselves when they get to a point where we wish they'd just take the word of authority (meaning us). Probably it's wisest to accept that in the end we just have to let go. I meet a lot of people in Christian work who came through a crisis of doubt or who (like me) converted from agnosticism. I also meet a lot of people who are still in the church but were so put off by a rigid religious upbringing that they have very little sales resistance to any liberalizing idea that turns up.

Like it or not, our world is a marketplace for good and bad ideas just as Athens was in Paul's day. The media and the Internet ensure that no reservation is sealed off from those ideas. We can do our best to prepare young people for what is coming, and to protect them for a little while, but it in the end they have to go out into that marketplace by themselves.

There is no guarantee that freedom of inquiry will generate the answers we want—that's why we call it freedom! This bothers a lot of people who don't want to participate in a search for truth unless they are assured in advance that the truth will be one they can accept. What if the challenge to materialism fizzles once again, as it did at the real Scopes trial? Many accommodationists in the Christian academic world, and some fundamentalists, have warned me that it is futile and dangerous to challenge the truth claims of modernism on secular territory. (To continue with the play as our metaphor, they are playing Rachel Brown to my Bert Cates.) I try to reassure them that today's scientific materialists are as overconfident as Matthew Harrison Brady was in the play, but there is no way to find out for sure without going into the courtroom.

To be sure, the risk is real. Modernists may insist on occupying the whole realm of reason, but they do leave a protected reservation for religious beliefs based on "faith." That reservation is our one "talent." Like the timid servant in Jesus' parable, we may be afraid of losing everything if we risk it in the marketplace of reason.

This is another one of those forks in the road. As long as the secular intellectual world is irrevocably committed to materialism, then Christian doctrines like supernatural creation and the resurrection are false by definition and can hardly survive academic scrutiny. Conversely, if those doctrines are true, then materialism, as a general worldview, *isn't* true. In that case the rules of the secular academy are open to question, to put it mildly. To step off the reservation to question the rules of the larger society is to take a great risk, but perhaps also to find a great opportunity. We will never know how great the opportunity was if we are afraid to take the risk.

When I drive from my home to my office at the University of California, I often pass Berkeley's public high school. Its façade displays in bold capitals one of the world's most familiar quotations, without (of course) identifying the source or the context: "YOU SHALL KNOW THE TRUTH, AND THE TRUTH SHALL MAKE YOU FREE." If you reread the context of that statement (the Gospel of John chapter 8), you will see why I always smile when I see it inscribed on one of our public schools. Our schools teach that what makes us free is knowledge, or political democracy, or the money and space to do our own thing. That's not the meaning of the quotation, not at all.

What is the Truth (with a capital *T*), and what does it set us free from? The rationalist philosophy of Henry Drummond in *Inherit the Wind* says that the source of Truth is science and its foundation is materialism. This kind of Truth sets us free from oppressive religious leaders who want to manipulate us for their own purposes. It also discredits the notion that there is a God who cares about what we do. Materialism sets us free from sin—by proving that there is no such thing as sin. There's just antisocial behavior, which we can control with measures like laws and educational programs. Why shouldn't that be enough?

Jesus was just as harsh in condemning religious hypocrites as Henry Drummond was, but he also warned against the scoffers who build their house on a foundation of sand. The Truth he referred to was himself, and the burden it frees us from is the sin that takes us away from our right relationship with the Father. Jesus said there is a way to be freed from our sins, but we can hardly find it if we aren't looking in the right direction.

That's the issue we need to open up for discussion. Our materialist blinders permit us to look in only one direction. Are we sure it's the right one?

Notes

Chapter 1: Emilio's Letter
The complete text of the 1995 Statement of the National Association
of Biology Teachers can be obtained from the NABT, 11250 Roger
Bacon Drive #19, Reston, VA 22090. Telephones: 800-406-0775; 703-
471-1134. The complete text appeared in *The American Biology Teacher*
58, no. 1 (January 1996): 61-62, and in the collection *Voices for Evolution*
(Berkeley, Calif.: National Center for Science Education, 1995), pp.
140-44.

The NABT statement also says that "evolutionary theory, indeed all
of science, is necessarily silent on religion and neither refutes nor
supports the existence of a deity or deities." This disclaimer does not
explain why the basic doctrine—that an unsupervised and impersonal
evolutionary process is our true creator—does not tend to refute the
existence of a deity. Probably the emphasis should be on the word
existence. Evolutionary science will allow God to "exist," if existing is the
only thing God ever does.

The quotation by George Gaylord Simpson is from his book *The
Meaning of Evolution*, rev. ed. (New Haven, Conn.: Yale University Press,
1967), pp. 344-45.

Chapter 2: Inherit the Wind
The script of the play *Inherit the Wind*, by Jerome Lawrence and Robert
Edwin Lee, was published in 1955 by Random House. Quotations are
from this version of the play; the movie dialogue differs in some details.
The movie makes clear that Bert Cates was teaching from Darwin's *The
Descent of Man*, not the more famous *On the Origin of Species* (which does
not deal with the evolution of humans from apes). Ironically, *The*

Descent of Man would never be allowed in a public school classroom today—because of its racism and sexism! For example, Darwin calmly predicted, "At some future period, not very distant as measured by centuries, the civilised races of man will almost certainly exterminate and replace the savage races" (*The Descent of Man* [Princeton, N.J.: Princeton University Press, 1981], p. 201). Imagine Henry Drummond trying to convince a modern jury that freedom of thought requires a community to accept the teaching of racial inferiority or genocide if it comes supported by "science."

For the most complete historical account of the Scopes trial in context, see Edward J. Larson's *Summer for the Gods: The Scopes Trial and the Continuing Evolution Debate* (New York: BasicBooks, 1997). For a shorter treatment of the difference between the play and the historical reality, see Carol Iannone, "The Truth About *Inherit the Wind*," *First Things,* February 1997, p. 28. Articles from *First Things* are available on the Web at http://www.firstthings.com.

Henry Fairfield Osborn, head of the American Museum of Natural History at the time of the Scopes trial, was the leading public antagonist of William Jennings Bryan, although he did not go to Dayton for the trial. Osborn was a fervent supporter of the discredited Nebraska Man and Piltdown Man fossils as proofs of evolution. Stephen Jay Gould has written engaging essays about Bryan and Osborn. See "William Jennings Bryan's Last Campaign" and "An Essay on a Pig Roast" in the collection of Gould essays *Bully for Brontosaurus* (New York: W. W. Norton, 1992).

A friend who read this chapter in manuscript provided the following revealing paragraph from Osborn's 1925 book (1925 was the year of the Scopes trial) *The Origin and Evolution of Life:* "In contrast to the unity of opinion on the *law* of evolution is the wide diversity of opinion on the *causes* of evolution. In fact, the causes of the evolution of life are as mysterious as the law of evolution is certain. Some contend that we already know the chief causes of evolution, others contend that we know little or nothing of them. In this open court of conjecture, of hypothesis, of more or less heated controversy, the great names of Lamarck, of Darwin, of Weismann figure prominently as leaders of different schools of opinion; while there are others, like myself, who for various reasons belong to no school, and are as agnostic about Lamarckism as they are about Darwinism or Weismannism, or the more recent forms of Darwinism, termed Mutation by de Vries. In truth, from the period of the earliest stages of Greek thought man has been eager to discover some natural cause of evolution, and to abandon the idea of supernatural intervention in the order of nature. Between the appearance of *The Origin of Species,* in 1859, and the present time there have been great waves of faith in one explanation and then in another: each of these waves of confidence has ended in disappointment, until finally we have reached a stage of very general scepticism" (H. F. Osborn, *The Origin and Evolution of Life* [New York: Scribner's, 1925], pp. ix-x).

As of the 1920s, it was difficult to say precisely what the "theory of evolution" was. Darwin's mechanism of natural selection was temporarily in eclipse, although his name had become practically synonymous with evolution. The triumph of neo-Darwinism (which Osborn calls Weissmanism) was in the future. The remarkable thing is that Osborn's summary is pretty accurate again today. The evolutionary scientists all agree that something called "evolution" is responsible for the history of life, but in professional circles the mechanism is once again up in the air. (For details, see chapter 4 of Phillip Johnson, *Reason in the Balance: The Case Against Naturalism in Science, Law and Education* [Downers Grove, Ill.: InterVarsity Press, 1995.])

Osborn even admits that the point of the whole evolutionary project, from ancient times to the present, has been to "abandon the idea of supernatural intervention in the order of nature." Evolutionary naturalists desperately want to believe they can put God away at a safe distance, and so "there have been great waves of faith in one explanation and then in another." Because the basic objective of denying the reality of God as Creator is contrary to reality, "each of these waves of confidence has ended in disappointment."

The Danny Phillips story was covered in a number of Denver-area newspaper stories and on radio and television shows. See especially Janet Bingham, "Boy Crusades Alone; Evolution Research Won Panel's Respect," *The Denver Post,* August 3, 1996, p. B1; and Sue O'Brien, "Zealots Rage from Left, Too," *The Denver Post,* August 18, 1996, p. F1. Eugenie Scott made her quoted comment on CBS-TV's program *Sunday Morning* on September 22, 1996. I learned of the outcome of the challenge directly from Danny Phillips and his volunteer attorney.

According to a teacher committee's report in Danny Phillips's school district, the complete text of the introduction to the NOVA video *The Miracle of Life* was as follows:

> Four and a half billion years ago, the young planet Earth was a mass of cosmic dust and particles. It was almost completely engulfed by the shallow primordial seas. Powerful winds gathered random molecules from the atmosphere. Some were deposited in the seas. Tides and currents swept the molecules together. And somewhere in this ancient ocean the miracle of life began. . . .
> The first organized form of primitive life was a tiny protozoan. Millions of protozoa populated the ancient seas. These early organisms were completely self-sufficient in their sea-water world. They moved about their aquatic environment feeding on bacteria and other organisms. They were covered with hundreds of tiny whipping hairs called cilia and flagella that made movement possible. From these one-celled organisms evolved all life on earth. And the foundation of life, the cell, has endured unchanged since the first tiny organisms swam in the cradle of life, the sea.

The teacher committee dryly observed that "a non-scientist might object to the statement that 'the first organized form of primitive life was a tiny protozoan . . . feeding on bacteria and other organisms' for reasons of belief, while a scientist might question how the protozoan could be the first form of primitive life if there were already bacteria to eat." The cilia and flagella that assemble themselves so easily in the above description are among Michael Behe's examples of irreducibly complex structures (see chapter 5), and scientists actually have no idea how they could have evolved.

Misleading and dogmatic statements are common in PBS NOVA programs on evolution, the producers apparently being more concerned to promote naturalistic philosophy than to portray the scientific uncertainties accurately. In early 1997 I participated in an Internet debate with Brown University biology professor Kenneth Miller in connection with the PBS NOVA television show *The Ultimate Journey*. This documentary featured photographs by Lennart Nilsson of human embryos developing in the womb. The accompanying narration labored mightily to insinuate the long-discredited doctrine that "ontogeny recapitulates phylogeny"—that is, that the embryo goes through a series of animal stages corresponding to the supposed evolutionary history of the species. Professor Miller did not defend the program but tried to change the subject to talk about hominid fossils and other stock arguments for Darwinism. Our written debate may still be available at the PBS/NOVA website http://www.pbs.org/nova

Chapter 3: Tuning Up Your Baloney Detector
Carl Sagan's *The Demon-Haunted World: Science as a Candle in the Dark* was published by Random House (New York) in 1996. His essay "The Fine Art of Baloney Detection" makes up chapter 12 of the book, pp. 205-20. Sagan's paragraph attacking "people who are offended by evolution" appears on p. 327.

Richard Feynman's 1974 commencement Lecture at the California Institute of Technology is reprinted as "Cargo Cult Science" in the collection *Surely You're Joking, Mr. Feynman: Adventures of a Curious Character* (New York: Bantam, 1989), pp. 308-17.

The account of Mark Wisniewski and the Lakewood, Ohio, incident is taken primarily from the article "Creation Science Banned from Lakewood, Ohio, Classrooms," *Skeptical Inquirer,* January/February 1997, pp. 6-8. A story by Ulysses Torassa in *The Plain Dealer* (Cleveland), June 4, 1996, p. 1B, reports, "Among other things, Wisniewski had passed out creationist articles and told students to compare their ideas with evolutionists'. After completing the assignment, half of the students said the creationist positions were more plausible. Wisniewski also asked students to outline their personal beliefs on such issues as "the ultimate nature of man, the nature and solutions to evil and suffering and what happens after death."

The essay by National Academy of Sciences President Bruce Alberts, "Evolution Versus Creationism: Don't Pit Science Against Religion," was published in *The Denver Post,* September 10, 1996, p. B9. The essay is a compendium of the usual spin-doctor arguments that official science organizations rely on to stop any serious questioning of evolution or materialism before it can get started. I recommend that teachers look for essays of this kind and use them for critical-thinking exercises after students have read chapters three, four and five of this book. One thing to notice right away is the title: the debate is set up as pitting creation*ism* (that is, an ideology) against evolution (no *ism*, therefore a fact). No matter what the evidence may be, an ideology (especially a *religious* ideology) can never beat a "fact" in a debate conducted under scientific rules. Scientific materialists actually see the issue that way, and so they naturally frame the debate in those terms. I always insist that an *ism* be put on both words or neither. Let the debate be between the competing facts (creation and evolution) or the competing ideologies (creationism and evolutionism). Better still, let it be between theism and materialism. What was present and active in the beginning, God or matter? That frames the question correctly and levels the playing field.

Chapter 4: A Real Education in Evolution

The Supreme Court decision described in the second paragraph is *Aguillard* v. *Edwards*, 482 U.S. 578 (1987). The Justices probably did not mean to lay down a rule that the official theory of evolution may not be criticized or questioned in public school classrooms, but that was the effect of their decision. The Justices who signed the majority opinion seem to have been fooled by arguments from the science establishment that every claim made by the scientific elite about "evolution" is a matter of neutral fact and that all opposition to materialism comes from people who want to read the Bible to students instead of teaching them science. Perhaps a Justice who drives home in the evening from the Court will by now have noticed the "Darwin fish" bumper stickers on cars—showing a fish with legs in mockery of the Christian fish symbol on other cars—and will realize that the Supreme Court has been duped into taking sides in a religious debate.

The cases in the lower federal and state courts invariably uphold disciplinary action against teachers who assert the validity of creation as an alternative to evolutionary naturalism. Any law library can supply a compilation of the cases under this citation: 102 A.L.R. Fed. 537 (1996). The law will change only when the courts become aware that there are genuine intellectual challenges to materialism and evolutionary naturalism. That is why Christians must be confident that they understand how to avoid being confined in the *Inherit the Wind* stereotype before they venture to argue the issues in public.

The quotation from Charles Darwin is from *On the Origin of Species* (New York: Penguin, 1982), p. 66. The quotation from Niles Eldredge about how evolution "never seems to happen" is from his book *Reinventing Darwin: The Great Debate at the High Table of Evolutionary Theory* (New York: John Wiley & Sons, 1995), p. 95. I have often wondered how Niles Eldredge and Steven Jay Gould can come so close to repudiating Darwinism outright without realizing what they are doing. I think the answer must be that materialism has taken hold so deeply in their minds that they do not understand that it is extremely vulnerable to criticism if the "blind watchmaker" mechanism is discredited.

Professor Tim Berra's remarks about the evolution of the Corvette automobile, with accompanying photographs, may be found on pp. 118-19 of his book *Evolution and the Myth of Creationism: A Guide to the Facts in the Evolution Debate* (Stanford, Calif.: Stanford University Press, 1990). I was at first stunned to learn that many evolutionary scientists do not understand the difference between common design and naturalistic evolution, even after I have explained it to them. A related misunderstanding is their tendency to cite embryonic development (the growth of the fetus in the womb) as an example of "evolution." Embryonic development is a programmed process that proceeds directly to a preordained end point. The apparent impossibility of using chance mutations to alter embryonic development so as to produce a different kind of animal argues strongly against the claims for Darwinian macroevolution. Such well-documented findings of embryology are invisible to persons whose minds are controlled by materialist philosophy.

Francis Crick's paragraph advising the public to read Dawkins "to save your soul" is from his book *What Mad Pursuit: A Personal View of Scientific Discovery* (New York: BasicBooks, 1988), p. 29. Richard Dawkins's words about nature's lack of interest in suffering come from *River out of Eden* (New York: BasicBooks, 1995), p. 131.

Chapter 5: Intelligent Design
The quotations by George Williams are from his interview in *The Third Culture: Beyond the Scientific Revolution*, ed. John Brockman (New York: Simon & Schuster, 1995), p. 42-43. See also George C. Williams, *Natural Selection: Domains, Levels and Challenges* (New York: Oxford University Press, 1992). The quotations from Richard Dawkins near the beginning of the chapter are from his books *River out of Eden* (New York: BasicBooks, 1995), p. 17 ("killing blow to vitalism"), and the preface to the 1976 edition of *The Selfish Gene* (Oxford: Oxford University Press, 1989), p. v ("robot vehicles blindly programmed"). These remarks are typical for Dawkins, who is only secondarily a biologist and primarily an evangelist for atheism and materialism.

I sent a preliminary version of my analysis of the matter-information problem to an academic journal called *Biology and Philosophy*, edited by Michael Ruse, hoping to draw a response from Williams and Dawkins. My paper, titled "Is Genetic Information Reducible?" was published in the October 1996 (vol. 11) issue of that journal (which appeared belatedly in February 1997). Dawkins and Williams did reply. My essay is on pp. 535-38; their replies are on pp. 539-41. Both of them state, correctly, that the problem of accounting for the origin of the information is not difficult if the information content of the organism is sufficiently low. Dawkins imagines a case of a "hypothetical book of nonsense character strings" (p. 540). Williams observes that "the pattern of slow-moving waves in sand dunes records information about what the wind has been doing lately. . . . The only author recognizable here is the wind" (p. 541). True enough, but the wind does not produce the kind of highly specified information required for a book or computer program or organism. Williams then gets to the crucial point, arguing that "the author of genetic information is as stupid as the wind" because, in Williams's opinion, animal bodies incorporate certain "functionally stupid historical constraints." That's the issue, all right. Does it require no more intelligence than the wind possesses to write *Don Quixote* or Windows 95—or to specify the genetic information required to create Miguel Cervantes or Bill Gates?

The issues of intelligent selection (Berra's Blunder) are brilliantly discussed in David Berlinski, "The Deniable Darwin," *Commentary,* June 1996, and especially in the follow-up symposium "Denying Darwin: David Berlinski and Critics," *Commentary,* September 1996, pp. 4-39. Publication of this article and symposium was particularly significant because *Commentary,* published by the American Jewish Community, had previously shown no interest in challenging the neo-Darwinian theory.

The quotation in the footnote is from an unsigned review of Michael Behe's book (see below) in *Skeptic* 4, no. 3 (1996), available on the Web at http://www.spacelab.net/~catalj/box/skeptic.htm.

Michael J. Behe's book is *Darwin's Black Box: The Biochemical Challenge to Evolution* (New York: Free Press, 1996). The long quotation is from pp. 18-20, where it comes with an illustration. The book reviews discussed in this chapter are James A. Shapiro, "In the Details . . . What?" *National Review,* September 19, 1996, pp. 62-65; Jerry Coyne,

"God in the Details," *Nature* 383 (September 19, 1996): 227-28. My own review of the Behe book and Richard Dawkins's *Climbing Mount Improbable* (New York: Viking, 1996) was published in *First Things,* October 1996, pp. 46-51, with the title "The Storyteller and the Scientist." This review is available for downloading at the Access Research Network Website.

The long quotation by Dawkins ("Physics books may be complicated ...") is from his earlier book *The Blind Watchmaker* (London: Longman, 1986), pp. 2-3. Dawkins at least has the issue right (where did that apparently irreducible complexity come from?), although his answers are made up of about nine parts imagination to one part fact. See also the reviews of Behe and Dawkins, and the resulting symposium involving many leading players in the debate, in the January and February 1997 issues of *The Boston Review,* available on the Web at http://www-polisci.mit.edu/BostonReview.

Richard Lewontin's comments about the a priori commitment of materialists to the philosophy (materialism) over the science come from his review of Sagan's *The Demon-Haunted World* in *The New York Review of Books,* January 9, 1997, pp. 28, 31. The conversation between David Chalmers and Kristof Koch is from John Horgan, *The End of Science: Facing the Limits of Science in the Twilight of the Scientific Age* (Reading, Mass.: Addison-Wesley, 1996), pp. 181-82.

Chapter 6: The Wedge

The complete text of the pope's statement is reprinted in *First Things,* March 1997, pp. 28-29. To summarize, the pope said that the theory of evolution is "more than a hypothesis" because it has been supported by several independent lines of research (unspecified); that there is more than one theory of evolution rather than a single theory; and that materialist theories of evolution are contrary to the church's teachings about the nature of humankind. The most authoritative commentary known to me is that of Cardinal Thomas J. Winning of Scotland, who wrote in *The Glasgow Herald* (January 11, 1997, p. 19) that the consistent position of the Catholic Church has been that "the Church leaves the believer free to accept or reject the various evolutionary hypotheses so long as they do not insist that the mind and spirit of man simply emerged from the forces of living matter with no room for God." The church has been concerned to state the minimum requirements of Catholic theology and has left scientific issues to the scientists.

Advanced students may want to try out their critical thinking skills on Stephen Jay Gould's treatment of the pope's statement in his essay "Nonoverlapping Magisteria" in *Natural History,* March 1997, p. 16. Gould exploits the *Inherit the Wind* stereotype, and the ambiguity in the term *evolution,* to further his argument that "science and religion are not in conflict, for their teachings occupy distinctly different domains." In effect, he is urging his readers (many of whom may be religiously inclined) to commit what we identified in chapter one as Emilio's third mistake.

The videotape of the 1994 debate at Stanford University between myself and William Provine is available from Access Research Network under the title "Darwinism: Science or Naturalistic Philosophy?" A study guide is available to accompany this video. It is highly suitable for use as a supplement to this book for instructional purposes. To obtain this and other video- and audiotapes, telephone ARN at 719-633-1772. E-mail: arn@arn.org Web: http://www.arn.org/arn

The Richard Rorty quote is from his very stimulating review of Paul Feyerabend's autobiography in *The New Republic,* July 31, 1995, pp. 35-36.

The papers at the 1992 Symposium at Southern Methodist University, including Michael Behe's first effort, were subsequently published in the collection *Darwinism: Science or Philosophy?* ed. Jon Buell and Virginia Hearn (Dallas: Foundation for Thought and Ethics, 1994).

Chapter 7: Modernism
Stanley Miller's 1953 experiment and the present state of origin-of-life research is discussed in chapter 8 of Phillip Johnson, *Darwin on Trial,* 2nd ed. (Downers Grove, Ill.: InterVarsity Press, 1993). The papers and speeches at the 1959 Chicago Darwin Centennial were published in three volumes under the title *Evolution After Darwin,* ed. Sol Tax (Chicago: University of Chicago Press, 1960). Julian Huxley's remarks are from his lecture published in the third volume of this set.

The Supreme Court decision in the school prayer case is *Engel* v. *Vitale,* 370 U.S. 421 (1962). In *Wallace* v. *Jaffree,* 472 U.S. 38 1985), the Supreme Court held unconstitutional an Alabama statute authorizing a one-minute period of silence in public schools "for meditation or voluntary prayer," on the ground that the purpose of the statute was to endorse religion. These decisions continue to spark controversy. Nearly all legal experts agree that the public schools should be in some sense "neutral" on religious questions, but many critics of the decisions argue that a rigidly secularized public education is far from neutral. The schools purport to teach practically everything students need to learn, from the "three R's" to sex education and driver training, and God evidently is not one of the things public educators think students need to know about. Although I agree with the critics that the public schools have effectively endorsed agnostic rationalism as the established civil religion, with naturalistic evolution as the creator, I do not think that this unbalanced situation can be changed by political or legal action. The underlying cause of the legal situation is the domination of the intellectual world by naturalistic philosophy, and this will not change until the ideas change.

The Supreme Court decision holding unconstitutional laws based on "animosity" toward homosexuality is *Romer* v. *Evans,* 116 S.Ct. 1620 (1996). The law in question was a Colorado state constitutional provision, passed by voter referendum, which forbade state or local governing bodies to pass laws forbidding discrimination against persons on the basis of their "homosexual, lesbian or bisexual orientation, conduct, practice, or relationships." The Supreme Court majority opinion by Justice Anthony Kennedy (a Reagan appointee) held that the law unfairly burdened the right of gays and lesbians to seek laws protecting themselves from discrimination and that the law lacked a rational basis because it was evidently "born of animosity toward the class that it affects." Justice Antonin Scalia's dissent characterized the law as "a modest attempt by seemingly tolerant Coloradans to preserve traditional sexual mores against the efforts of a politically powerful minority to revise those mores through use of the laws." The underlying philosophical and religious question is whether "traditional sexual mores" are themselves unconstitutional insofar as they disfavor same-sex relationships.

Chapter 8: Stepping off the Reservation
The account of Charles Templeton, Billy Graham and John Shelby Spong is take from the *Time* magazine cover story on Graham in the issue of November 15, 1993, which is based on the William Martin biography.

I am encouraging educational institutions, particularly Christian colleges, to develop special curricula designed to prepare students to meet the intellectual challenges of evolutionary naturalism and to develop confidence in the intellectual strength of theism. One pilot program that I hope will serve as one model for doing this is the Torrey Honors Institute at Biola University in La Mirada, California. It is a Great Books type of curriculum with particular emphasis on the philosophical issues that underlie worldviews. Initial response to this program shows that there are a great many students with fine academic records and strong Christian commitment who are eager to master the intellectual skills needed to stand up to the naturalistic bias they will encounter in the secular culture. Probably this kind of teaching can be delivered in many different types of programs to meet individual needs and interests, once institutions and donors see the possibilities. For information on the Torrey Honors Program, contact Professor John Mark Reynolds at Biola Univerity, 13800 Biola Avenue, La Mirada, California 90639 (e-mail: johnr@isaac.biola.edu).